Taste, Politics, and Identities in Mexican Food

Also Available from Bloomsbury:

Cooking Technology, edited by Steffan Igor Ayora-Diaz
Making Taste Public, edited by Carole Counihan and Susanne Højlund

Taste, Politics, and Identities in Mexican Food

Edited by
Steffan Igor Ayora-Diaz

BLOOMSBURY ACADEMIC
LONDON • NEW YORK • OXFORD • NEW DELHI • SYDNEY

BLOOMSBURY ACADEMIC
Bloomsbury Publishing Plc
50 Bedford Square, London, WC1B 3DP, UK
1385 Broadway, New York, NY 10018, USA

BLOOMSBURY, BLOOMSBURY ACADEMIC and the Diana logo are
trademarks of Bloomsbury Publishing Plc

First published in Great Britain 2019

Cover image: constantgardener/Getty

A catalogue record for this book is available from the British Library.

A catalog record for this book is available from the Library of Congress.

ISBN: HB: 978-1-3500-6667-0
ePDF: 978-1-3500-6668-7
eBook: 978-1-3500-6669-4

Typeset by Integra Software Services Pvt. Ltd.
Printed and bound in Great Britain

Contents

Illustrations

Maps

Tables

Contributors

Steffan Igor Ayora-Diaz is Professor of Anthropology at the Autonomous University of Yucatán. Since 2008 he has led the research group Studies on Social Practices and Cultural Representations. He is a member of the Mexican National System of Researchers (Level II) and has been president of the Society for Latin American and Caribbean Anthropology, section of the American Anthropological Association (AAA). His publications include *Foodscapes, Foodfields and Identities in Yucatán* (2012), the coauthored book *Cooking, Music and Communication: Aesthetics and Technology in Contemporary Yucatán* [in Spanish] (2016), and *Cooking Technologies: Transformations in Culinary Practice in Mexico and Latin America* (2016).

Sarah Bak-Geller Corona is a professor at the Institute of Anthropological Research of the Universidad Nacional Autónoma de México, and a member of the National System of Researchers (Level I). Her research focuses on the politics of culinary practices and forms of representation within colonialism and nation-formation in Mexico and Latin America. She is part of the Mexican Group for the Anthropology of Food and the international project *Foodherit*. She is an associate member of the Laboratory Patrimoines Locaux of the National Museum of Natural History, France. Her publications include *To Inhabit the Kitchen* [in Spanish] (2006) and *Cuisine, Society and Politics in Mexico, 16th–19th Century* [in French] (In Press).

Ronda L. Brulotte is Associate Professor of Geography and Environmental Studies and Director of Latin American Studies at the University of New Mexico. She has conducted ethnographic fieldwork in Mexico since 1998, and is trained in the anthropology and cultural geography of Latin America more broadly. She is the author of *Between Art and Artifact: Archaeological Replicas and Cultural Production in Oaxaca, Mexico* (University of Texas Press 2012) and co-editor, with M. A. Di Giovine, of *Edible Identities: Food as Cultural Heritage* (Routledge 2014). She is currently working on a book manuscript addressing the transformation of the Oaxacan mezcal industry within the context of the global market.

Guadalupe Cámara Gutiérrez is a researcher investigating the production, commercialization, and consumption regulations of *aguardiente* during the nineteenth century in Yucatán. Among her recent publications there are several articles and book chapters: "Aguardiente and Transgression in Yucatán, 1875–1895," in *Norms, Transgressions. Order Infractions in Yucetecan Society*, edited by P. Miranda Ojeda and P. Zabala Aguirre (2009), and "Alambiques and Aguardiente Drinkers," in *Our History in Lowercase*, compiled by G. Negroe Sierra and P. Miranda Ojeda (2010).

Jeffrey H. Cohen is Professor of Anthropology at Ohio State University. His research on food in Oaxaca, Mexico, has appeared in *Gastronomica*. He is the author of many articles and books including *Cooperation and Community: Economy and Society in Oaxaca* (1999) and most recently *Eating Soup without a Spoon: Anthropological Theory and Method in the Real World* (2015).

Lilia Fernández Souza is a professor in the Faculty of Anthropological Sciences of the Autonomous University of Yucatán and coordinates the Chemical and Microscopic Analyses Laboratory at the Faculty. She is currently director of the project "Daily life at Sihó Yucatán: Social and economical diversity in non-elite households of a Classic community." She has edited the books *Vida cotidiana de los antiguos mayas del Norte de la Península de Yucatán*, with Rafael Cobos, and *En los Antiguos Reinos del Jaguar*. Her publications include: "Grinding and Cooking: An Approach to Mayan Culinary Technology" (2016) and "The Ancestral Stew Pot: Culinary Practices at a Contemporary Yucatecan Village" (2017).

Mario Fernández-Zarza is a PhD candidate in the Department of Social Anthropology at the Universidad de Sevilla. He is a member of the research group "Territory, Culture and Development" of the Universidad de Sevilla. His research topics include agro-food systems, local development, and new technologies and food culture.

Héctor Hernández Álvarez is a professor-researcher in the Facultad de Ciencias Antropológicas of the Universidad Autónoma de Yucatán and a member of the National System of Researchers (Level I). He is currently director of the project "San Pedro Cholul: geoarqueología, historia e industrialización de una hacienda henequenera yucateca de principios del siglo XX." He edited the book *Sendas del henequén: un estudio arqueológico de la hacienda San Pedro Cholul, Yucatán* (2016).

Ignacio López-Moreno is a professor in the Department of Social Processes of the University Autónoma Metropolitana and a member of the National System of Researchers (Level I). He coordinates the Mesoamerican Corn Observatory and the Research Network "Corn: Nutrition, Technology, Ecology and Culture," and is a member of the research group "Territory, Culture and Development" of the Universidad de Sevilla, and the research area "Public Policies, Economy, Society and Territory" at UAM. His publications include: *El Maíz Nativo en México. Una aproximación crítica desde los estudios rurales* (2016) and "Los alimentos de calidad. Nuevas estrategias rurales para nuevos consumidores," *Arxiu d'Etnografia de Catalunya* (2016).

Ramona L. Pérez is Professor of Anthropology and Director of the Center for Latin American studies at San Diego State University. She is Chair of the Institutional Review Board at San Diego State University and graduate faculty in Global Health (SDSU/UCSD), Women's Studies, and at the Colegio de la Frontera Norte (COLEF). Her current work focuses on binational youth identity and family composition, migration and the transmission of identity between mothers and their children, shifts in culinary food practices and nutrition, and the moral economy of lead poisoning in ceramic production. She has published in multiple journals in the fields of anthropology, geography, public health, social work, criminal justice, and medicine.

Jeffrey M. Pilcher is Professor of History and Food Studies at the University of Toronto. He is the author of several books on the history of Mexican food, including *¡Que vivan los tamales! Food and the Making of Mexican Identity* (1998); *The Sausage Rebellion: Public Health and Private Enterprise in Mexico City* (2006); and *Planet Taco: A Global History of Mexican Food* (2012). He also edited *The Oxford Handbook of Food History* (2012) and *Food History: Critical and Primary Sources* (2014). He is the editor of the peer-reviewed journal *Global Food History*.

Paulette Kershenovich Schuster graduated from The Hebrew University of Jerusalem where she also conducted research on Jewish Mexican women living in Israel as part of her post-doctorate. She is the author of a book on Syrian Jewish and Lebanese Maronite women in Mexico City. She has written more than twenty academic articles on religion and food, identity, and women in Latin America, among many other topics.

Gabriela Vargas-Cetina is Professor-Researcher of Anthropology in the Facultad de Ciencias Antropológicas, Universidad Autónoma de Yucatán. She has done fieldwork in Canada, Italy, and Mexico. She has been president of the Society for Latin American and Caribbean Anthropology, section of the AAA, and a member of the Executive Board of the same organization. Her work has centered on organizations, the arts, and representation practices in academic writing. Her latest book is *Beautiful Politics of Music: Trova in Yucatá, Mexico* (2017). Her current interests include the anthropology of technology and the anthropology of performance in the context of organizations, music, sound, and space.

Christine Vassallo-Oby is a lecturer and Director of Undergraduate Studies in the Department of Latin American, Caribbean, and US Latino Studies at the University at Albany, SUNY. Her research focuses on the culture, economy, and history of the cruise ship industry in Cozumel, Mexico, and other Caribbean port-of-call communities. Her recent publications include the book reviews, *Tourism and the Power of Otherness: Seductions of Difference* (2014) by D. Picard and M. Di Giovine, eds. in *Tourism: An International Interdisciplinary Journal* (2017) and *Imagined Globalization* (2014) by Néstor García Canclini in *Journal of Anthropological Research* (2015), and the article "Cruise Ship Tourism in Cozumel, Mexico: 'Frios Como la Naturaleza de los Gringos lo Dice'" (*SfAA News*, 2014).

Richard Wilk is Distinguished Professor and Provost Professor Emeritus at Indiana University where he co-manages the Indiana University Food Institute and continues to participate in an innovative PhD track in Food Anthropology. He has also taught at the University of California (Berkeley and Santa Cruz), New Mexico State University, and University College London. He has recently begun fieldwork in Singapore with a Fulbright teaching and research fellowship. His most recent books are *Exploring Everyday Life* (2016), coauthored by Orvar Lofgren and Billy Ehn, and a coedited collection with Candice Lowe Swift, *Teaching Food and Culture* (2015).

Acknowledgments

The inception of this volume is traceable to a small colloquium I organized to frame the inauguration of the Gustemology Laboratory at the Facuty of Anthropological Sciences, Autonomous University of Yucatán, in June 2016. Only five authors, with four papers (different from those presented at the event), are represented in this volume, Sarah Bak-Geller Corona, Héctor Álvarez Hernández and Gudalupe Cámara Gutiérrez, Lilia Fernández Souza, and Steffan Igor Ayora-Diaz. Still, the spirit of the presentations and dialogue with Miriam Bertram Vilà, Humberto Thomé-Ortiz, and Fernando Enseñat inspires some of the themes we discuss in this volume. The chair of the Faculty of Anthropological Sciences, Dr. Celia Rosado Avilés, relieved me of teaching obligations during the Spring 2018 term, to facilitate my task of completing this manuscript.

In addition, I would like to thank Lucy Carroll and Miriam Cantrell at Bloomsbury, for their interest in this volume and for providing very useful guidance throughout the process. Finally, my thanks to Cheryl Merritt for the fabulous line editing and setting of this book.

Introduction: Matters of Taste: The Politics of Food and Identity in Mexican Cuisines

Steffan Igor Ayora-Diaz
Universidad Autónoma de Yucatán

From flavor to taste

When Jean Anthelme Brillat-Savarin published *The Physiology of Taste* in the early nineteenth century, he had a more complex vision of the sources of taste than the one developed later during the twentieth and early twenty-first centuries. In the following decades, modern rationality spawned a rational-scientific logic with reductionist tendencies, which led scientists to privilege human biology over other aspects of the experience and multiple meanings of taste. In this volume we focus on Mexican food, not as a unitary tradition, concept, or set of practices but, rather, we show its diversity and question its unified construct because it hides, on the one hand, the political tensions and negotiations at work between the ideology of a purported *national* cuisine rooted in the indigenous past and, on the other hand, the actual local, ethnic, and regional cuisines in which hybridity thrives.

To understand taste we need to start by acknowledging the biological nature of the human body, but we believe that a productive critique of what taste means for different sociocultural groups must be achieved through the analysis of the social, economic, ecologic, and political context that shapes the availability of food. Here we develop an understanding of taste as those complex interactions involving biological, but also and more fundamentally, social, cultural, ecological, and political contexts that contribute to shape the human experience of food intake and preference. We seek to take advantage of theories, concepts, and methodologies that have emerged during the last three decades of the history of the humanities and social sciences in general, and anthropology in particular. These demand an examination and understanding of nonlineal, complex sociocultural and political processes that shape how people simultaneously develop shared and differing experiences of taste in food. This critique is even more necessary at this stage of our discipline when we have also become aware that both "nature" and "body" are politically and culturally contested terrains (e.g., Braidotti 1994; Descola [2005] 2013). Taste happens in the body, but it is experienced in a particular place (as locality) and time (as history), and in the specific economic and sociopolitical circumstances in which different groups give different meanings to their intersubjective experience of taste. During the nineteenth century, as the modern cultural condition settled in and shaped North Atlantic social and cultural understandings of human and nonhuman nature, scientific-rational discourse and practices engaged in the "purification" of the different sociocultural spheres (Latour 1983). This development has led to privileging the laboratory study of taste, i.e., the perception of flavors as an autonomous phenomenon to be studied in isolation from other empirical and theoretical domains. This book challenges these views reducing taste to a mere set of biological functions.

Flavor

Most research on taste, understood as the perception of flavors, has become the dominion of biologists, biochemists, physiologists, and psychologists. Consequently, there has been a long-lasting surge in this field of research focusing on the anatomical and physiological properties of the tongue and nose, the chemical properties of flavor and aroma stimulants, and the cognitive and emotional aspects tied to the experience of flavors. Those who have privileged the tongue have examined and discussed the spatial distribution of gustative papillae and its effects on the perception of four basic flavors: sweet, salty, sour, and bitter. Nonetheless, these scientists recognize that individuals and social groups may have different papillae distributions. Even from the standpoint of the biological sciences it is possible to argue that their experience of taste is determined by the configuration of the subjects' tongues, which in turn is probably determined genetically. Currently, new basic flavors are being added to the existing list: umami and fatty, for example—the list may continue to grow as different cooks, chefs, food journalists, corporations, and scientists claim new territories (Holmes 2017; McQuaid 2015; Spence 2017; Stuckey 2012).

In general, biologically informed research tends to reproduce the scientific, rational procedure of separating the different senses: taste/flavor, smell, touch, sight, and hearing are experienced through different organs. Lately, it has been argued that we need to distinguish between retro-nasal and nasal perceptions of aroma, as they inscribe different qualities in our experience of a food's taste (Shepherd 2012). Consequently, flavor may be first felt in the tongue constituting a very basic, simple experience. According to this line of reasoning, it is the perception of aromas that gives complexity to the taste of food. Taste scientists have reached the conclusion that it is through the interaction of the senses that we finally get to experience "taste," a more complex assemblage of sensations. Consequently, some authors have begun to look at sense integration (Shepherd 2012; Spence 2017). Along these lines, Mouritsen and Styrbæk (2017) pay attention to the part played by mouthfeel and argue that this sensation is in fact the result of the tactile and mechanical feel of the food in the mouth during mastication. However, they note that mouthfeel acts in concert with what we see, smell, and anticipate from the action of the other senses (see also Stillman 2002). Thus, they suggest that some flavors, which some authors have considered basic, such as stringent and fatty, are in effect the result of our mouthfeel perception—that is, the interaction between senses—and not flavors in themselves. Yet, these arguments about taste are all predicated on genetic (pre)dispositions, on the biology and physiology of the human body, in the chemical properties of food ingredients interacting with our organs of perception, and cognitive phenomena that help us establish what is good and what is bad tasting food. It can be argued that these approaches objectify and reify "taste," assuming it as a universally shared perception. Consequently, if taste resides in the body and in the objects it tastes, research can be focused on the development of new stimulants, on changes with age, or how genetics determines the likeability of an ingredient (cilantro, for example, that some individuals like and others dislike), or how different drugs modify our perception of one ingredient or another.

While in the biological sciences researchers may claim to do research for knowledge's sake (ignoring or concealing the use of their studies in the development of chemical and synthetic enhancers for the flavors, colors, and aromas of industrially processed food), in the social sciences and humanities we examine the different ways in which taste can be, and is, used for social, political, and economic purposes (Wilson 2013). A couple of seminal volumes on taste and the senses have successfully assembled essays from different disciplines illustrating the multiplicity of political, social, and economic agents and processes involved in the constitution and institution of what we generically call "taste," as well as in the experience of all the other senses (Howes 2005; Korsmeyer 2017). As the contributors to those volumes demonstrate, in addition to the biological substratum of taste we need to examine all the multiple dimensions that taste brings together.

The different chapters in this volume show how taste, and its different meanings, is socially and culturally negotiated and constituted. Subjective and intersubjective practices and discourses, memory and emotions, forms of commensality, the perception of identity and difference, are all grounds for the emergence of locality, translocality, ethnicity, regionalism, and nationalism as grounds for affirming cultural identities. These all have a cultural-political dimension that is reflected in shared tastes, which these collectively manifested taste preferences contribute to intensify. For example, as a reflection of a homogenizing nationalist discourse, growing numbers of people may decide to showcase their Mexican identity by choosing *chilaquiles* for breakfast, and the fact that they choose these, mole, and other recognizable "Mexican" meals while refusing other "national" meals, and despising those who eat them, contributes to confirm and intensify the politics of one of the possible modes of culinary nationalism.

Taste

Authors such as Agamben ([1979] 2017), Gronow (1997), and Korsmeyer (1999) have traced the transformation of the meaning of taste, from something that relates to our perception of flavors into an aesthetic system of values for the judgment of art. They show how, since the times of classic Greek philosophers, the perception of flavor underwent a process of devaluation where its individual, subjective nature was placed, along with the sensations of touch and smell, below sight and hearing. Sight and hearing are understood as senses that operate at a distance giving humans the possibility to be "objective." Immanuel Kant's philosophical arguments about the emergence of judgments of beauty and of aesthetic values supplemented and justified the devaluation of taste, smell, and touch on the grounds that the distance granted by sight and hearing allowed for the disinterested appreciation of the aesthetic properties of objects. The other three senses, in effect, were important in the pleasurable and interested appreciation of objects that permit the reproduction of the species: sex and food. Thus, for him and for those following on his trail, they were subjective, interested senses and poorly conducive to objective aesthetic judgment.

Within the social sciences, taste has been turned into a social marker used to distinguish between members of different groups with historically different access to economic, social, and cultural capital. As Bourdieu ([1979] 1984) illustrated in his ambitious analysis of the sources of social distinction, members of different groups are characterized by differences in their taste for music, literature, clothing, museums, art, and food. Taste is conditioned by social membership, and at the same time it marks that membership. This formulation has been taken, albeit in critical ways, by different authors (e.g., Gronow 1997; Korsmeyer 1999), and continues to influence how we understand the social and political meanings of taste—as different chapters in this volume show. Furthermore, as Bourdieu (1991) showed, taste is socially codified, and the ability to constitute and institute codes of taste is a manifestation of symbolic power within the political field. There are those who have been recognized with the authority to represent the other members of society, and they have the discursive instruments to formalize aesthetic codes. Along these lines, Irvine and Gal (2003) speak of "language ideologies" as a discursive configuration that allows a group of people to impose onto other people their frames of reference and their conceptual schemes, charged with moral and political values and contingent on the social position each group and individual occupies within a power structure (2003: 35). This is important, as I explain below, and it is suggested in several chapters when individuals and agencies seek to establish the definition of what *is* national Mexican cuisine.

Taste in and for food is thus a political matter. It becomes the realm in which identities are affirmed and differences are highlighted, where inclusion and exclusion happens. Taste plays an important part in defining class differences, including differences in cultural, social, and symbolic capital that families and groups have accumulated over time. Cultural capital, in the form of education and the acquisition of "culture," contributes to shape class differences, as shown by the essays in Strong's (2011) volume. There is nothing natural in *favoring* the taste of beer, whisky, fruits and salads, or meat, or, more basically, sweet, sour, salty, or bitter. Our choice of restaurants, when we go out to eat, is easily turned into a marker of social belonging, class, and exclusion (Finn 2017; Lane 2014). Within this system of class differences mediated and motivated by food, so-called foodies of different types can affirm their identities and differences through their selection of meals, ingredients, and restaurants (Johnston and Baumann 2010).

The politics of taste can be revealed both in the private and public domains—although we are all aware that the boundaries between the private and the public are always permeable, and hence there is always the possibility to negotiate meanings and practices through taste displayed in different contexts. The often-cited chapter by Stoller and Olkes (1989) illustrates how intentionally altering the ingredients and flavor of a sauce can expose intrafamilial and interethnic tensions. Gender and religious differences can also express conflicts within the domestic realm (Appadurai 1981), and the household table can be turned into a site for the display of family disharmony despite views romanticizing family commensality (Wilk 2010). Still, families and groups of friends share foods in different social contexts. These experiences are inscribed in the biographies of individuals and groups, and they ground the shared memory of meals attached to places, times, people, and individuals (Sutton 2001). Taste, as Proust so famously illustrated, is rooted in our memory and in nostalgia, and we can judge the quality of the taste of what we eat on the basis of those past experiences. Taste is predicated upon our memories of past foods and the repetition of ingredients, meals, and commensality on an everyday basis (Ayora-Diaz 2012a, 2012b). This sociocultural negotiation of the meaning of taste roots local, ethnic, regional, and national identities. These, as it has been often suggested, are conditioned both by discourses fostered by homogenizing nation-states and the everyday practices that support the existence of a "popular" nationalism (Edensor 2012), regionalism (Ayora-Diaz 2012a), and other forms of localism.

This assertion would resonate with the position taken by different authors dealing with "national" cuisines—for example, in France (Ferguson 2004), India (Appadurai 1988), Italy (Camporesi 1970; Capatti and Montanari 1999), Japan (Cwiertka 2006), and Mexico (Pilcher 1998), to name a few. As these authors have demonstrated, those in political power seek to promote nationalistic feelings, and whenever food is a meaningful marker of distinction and difference, cuisines are made into privileged symbols of difference. However, these pedagogical impulses always meet, albeit with different intensity, performative forces that at the everyday level can reproduce state visions or challenge them to different extents (Bhabha 1994). This relationship of political forces reveals the long-lasting tensions between homogenizing forces and local and translocal movements that lead to fragmentation and multiplicity, challenging the notion of one single imagined community.

Is there a "Mexican" cuisine (and taste)?

Regarding the politics of taste in Mexican food it is probably best to advance a seemingly controversial statement, one that highlights the tensions manifest throughout this volume's chapters: there is no such a thing as "Mexican food." Mexico encompasses a large territory in which about 85 percent is mountainous and contains deserts, valleys, highland plateaus, flat territories, dry and wet lands, and the seas bathing its coasts are the Caribbean, the Pacific, and the Gulf of Mexico. This diversity of landscapes creates different opportunities for agriculture, fishing, hunting, gathering, and for favoring or preventing the flows of edible commercial goods. Mexico's demography is also diverse. Each state encompasses from one to several aboriginal groups, and there are currently sixty-eight officially recognized Mexican indigenous languages (see National Institute of Geography and Statistics [INEGI] 2015), and each region has attracted immigrants from different regions of Spain, France, Germany, Italy, different east European countries, China, Korea, the Middle East, and more recently from Canada, the United States, as well as many different countries from all over the world. Additionally, aboriginal and immigrant groups show a tendency to move (despite the Mexican state's attempts to control their movements), and each region today hosts people from other Mexican states, who speak different languages and, important for this volume, like to eat different things, enjoying different flavors, aromas, colors, and textures. Within this multicultural context, despite recurring efforts to invent it, it is very difficult to maintain that there is *one* Mexican cuisine other than in institutional, state, and bureaucratic discourse.

Marked by particular and specific historical trajectories, each region has developed its own cuisine and its own taste based on a mixture of local ingredients and whatever the market allowed it to introduce at different times. Ingredients may be shared across groups, but the way and proportions in which they are mixed, the techniques used to prepare meals, and the technologies employed in their elaboration all result in a multiplicity of tastes, and each becomes meaningful in a different region. For example, the food of the states of Tamaulipas, Veracruz, Tabasco, Campeche, and Yucatán, all bathed by the Gulf of Mexico, each have their own taste and are distinguishable from each other, even if historically they have come to share multiple ingredients and cooking techniques.

This short volume does not cover all regional foods of Mexico.[1] The authors here are interested in how the taste of certain ingredients and the formal recognition of different culinary styles become instrumental to affirming local, regional, and national identities. Hence, we are trying to look beyond the nutritional and biological aspects of food, and instead look at the politics of taste in Mexican foods. This happens in a context in which there is a strong nationalist discourse supported by a power structure that has enabled the construction and dissemination of the notion that there is one single Mexican cuisine, one that erases cultural culinary differences in favor of a homogenizing and hegemonic culinary construct. Pilcher (1998) and Ayora-Diaz (2010, 2012a) have shown at the national and regional levels how the dissemination of this nationalist viewpoint contributes to erase or silence local and regional culinary differences. Hence, several chapters in this volume recognize the direct and/or indirect effects that the inclusion of the Michoacán paradigm by the United Nations Educational, Scientific and Cultural Organization (UNESCO) in the list of "Humanity's Intangible Cultural Heritage" has on regional cuisines and on the global market of ethnic and national cuisines under the label of "Mexican."

Nationalism and the Michoacán paradigm

In 2010, UNESCO included the Michoacán paradigm in its list of "Humanity's Intangible Cultural Heritage." The text on UNESCO's website is revealing of the political discourse deployed in Mexico to promote the imagination of a single Mexican cuisine: according to this document, the traditional Mexican cuisine comprises age-old skills, and "ancestral" community customs, based on the trinity of corn, beans, and chili peppers; the indigenous traditional cooks of Michoacán make tamales, as people do in all other parts of Mexico; and there exist collectives of traditional cooks (see UNESCO 2010). The Conservatory of Mexican Gastronomic Culture, with the support of state agencies (tourism and culture), and some Mexican food corporations, were able to achieve this recognition after a failed attempt in 2004 (Moncusí and Santamarina 2008).

The inclusion of the Michoacán paradigm in the UNESCO list is important as it has had several effects: first, the Conservatory of Mexican Gastronomic Culture has gained political ascendancy through its collaboration with the governments of different Mexican states to create collectives of "traditional" cooks. In fact, in 2018, several of these collectives are celebrating their first or second meetings—revealing their recent creation. Second, members of the conservatory visit the different Mexican states to collect "traditional" recipes in which corn, beans, and chili peppers are the main ingredients. On these grounds they are able to declare that the food of every state is "Mexican," silencing all the regional differences in culinary practices. Third, worried about the dangers that Mexican food faces abroad, in 2015 the conservatory proposed, along with the Mexican Senate, the creation of a certification procedure for restaurants of Mexican food on foreign soil. In 2016 the Mexican Senate promulgated the Law for the Advancement of Mexican Gastronomy, which included measures to protect Mexican cuisine within Mexico and abroad (Senado de la República Mexicana 2016). Fourth, its representatives recently signed an agreement with the Ministry of Education in which they became the official institution to grant recognition and official certification to "traditional" cooks (*20Minutos* 2018); and fifth, in response to this homogenizing tendency, some states are declaring their own cuisine the cultural heritage of the people who live in them. This has happened in Veracruz, Puebla, and Yucatán. In Oaxaca, it has been the state's culture that has been declared cultural patrimony and food plays an important part of the whole scenario affirming local and regional cultural identity (see Ayora-Diaz, this volume).

Yet, in every instance, as Foster (2015) has pointed out, the "local" is not univocal. Everywhere in Mexico, different groups compete with each other seeking to advance their own interpretation and choice of the practices, objects, or ideas they see as constituting their "true" local heritage. However, even in this context we can witness what Michael Hechter (1975) called "internal colonialism," as some versions of the local seek to obfuscate their own relative standing against other local forms of food. The construction of the modern nation-state is predicated, very often, in the erasure of regional differences, but these also erase, very often, competing ones. The Michoacán paradigm, in practice, is becoming instrumental in the imagination of a national cuisine that, in the twenty-first century, adopts the values of early twentieth-century nationalism. The imagination of a reductionist national cuisine based on beans, corn, and chili peppers erases the diversity of tastes within the Mexican territory.

The politics of taste in Mexican cuisines

As suggested above, given the heterogeneous landscape of the territory encompassed by the Mexican republic, each region has developed through history its own commercial networks; the inhabitants of each state have favored different migration targets; and each state has become in turn the target for different immigrant groups. The movement of people and edible commodities, different for each region, in addition to the ingredients that can be locally produced, favor through time the development of different tastes that become effectively tied to what each group of people consider their own cuisine, culture, and identity. For example, as I have shown elsewhere, the state of Yucatán has a very diverse demographic and ethnic composition where, for the last 500 years, immigrants mainly from Spain, Germany, France, the former Ottoman Empire (today Lebanon, Palestine, and Syria), China, and Korea imposed themselves over a somewhat linguistically homogeneous indigenous population: the Maya of the peninsula of Yucatán. The assemblage of taste preferences for certain ingredients and culinary technologies made possible the creation of a Yucatecan cuisine and gastronomy different from other Mexican regions, and more like their Caribbean counterparts (Ayora-Diaz 2012a). One consequence has been the political erasure, until recently, of Maya cooking and also of the cooking legacy of other immigrants with a long history of residence in the state of Yucatán, such as the Chinese and Koreans, who remained politically subordinated in the regional power structure. People of Syrian-Lebanese ancestors, in contrast, keep many dishes of their "traditional" repertoire, have restaurants specializing in Lebanese food, have their own social clubs and organizations, and their own church. They have managed to preserve their own food preferences, albeit modified to suit the local taste that by now they have come to share.

In comparison, Oaxaca encompasses four main linguistic, ethnically different groups. Hence, there are several culinary traditions vying for the power to represent the cooking of all people in the state. Magaña González (2016) has shown how different culinary practices converge and enter into competition and negotiation to define the food of this Mexican state, and has discussed how the Zapotec-speakers have gained ascendancy, even though Mixtecs, Mixes, and Mazatecos seek to affirm their own cultural identities through food as well. In Oaxaca, differently from Yucatán, it is "indigenous" cuisines that claim recognition as the state's cuisine. Also in contrast, while Yucatán is flat, lacks rivers, and is bathed by the Gulf of Mexico, Oaxaca has valleys, mountains, rivers, and is bathed by the Pacific Ocean. Hence, the inhabitants of each region have developed their own taste preferences, even though the tourism industry has chosen a few dishes to represent each of these states' cuisine.

In the meantime, the Mexican highlands, and in particular Mexico City, have been the seat of power since before the arrival of the Spaniards. This condition has been turned into the source of centralist and nationalist policies and political forms of cultural representation of the nation as a whole. Its own particular history shows transformations in the regional economy that made it possible to gain access to edible goods from other parts of the country (Pilcher 2006). As Bak-Geller Corona (2013) has argued, nineteenth-century Mexican cookbooks displayed this tendency to incorporate recipes from elsewhere at the same time that they gave them a nationalist twist (see also Bak-Geller Corona, this volume).

This impulse has been translated into the tendency to erase the multiplicity of regional cuisines during the invention of a Mexican cuisine (Pilcher 1998).

In general, it can be argued, the different access that each Mexican region had to locally grown produce and imported edible commodities made possible the emergence of different tastes for food and drink. Some regions were more inclined to drinking rum, while in others emerged what today are large industries of tequila and mezcal, as well as many regional breweries. Each local industry sought to quench the thirst and please the palates of different populations (e.g., Gaytán 2017). And while these were the result of importing different technologies from Europe, along with Mexican migrants, Mexican foods have also moved in the opposite direction, reaching Europe and other parts of the world, sometimes contributing to reshape other cuisines (Long 1992) but often seeking to satisfy the taste for home in different places around the world (Pilcher 2012). Unfortunately, with the exception of the United States, few studies have been conducted on the food practices of Mexican migrants in Europe, the Middle East, Asia, and elsewhere (see Kershenovich Schuster, this volume).

Structure of the volume

The chapters in this volume seek to explore how taste is negotiated and how the politics of cultural identity plays an important part, along with history, nostalgia, memory, commensality, mobility, social class, and global processes—such as the rapid growth of the tourism industry, including gastronomic tourism—in the constitution and repetition of culinary practices related to different Mexican cuisines. In addition to this introduction and the Postface at the end, this book is divided into three parts, each one including four chapters.

Part One, "Tasting the Past in Mexican Foods" begins with Chapter 1 by Lilia Fernández Souza, "A Touch of Pre-Columbian Maya Flavor." She argues that it is possible and necessary to produce a *tasteful* archaeology through the articulation of zooarchaeological findings along with chemical analysis of samples of cookware and ethnoarchaeological techniques. She looks at findings in the Maya region of Yucatán seeking to elucidate possible differences in the diets of commoners and elites. She argues that this approach can lead to a more complex understanding of daily lives and sensations in the past. Chapter 2, "Gastronomy and the Origins of Republicanism in Mexico," by Sarah Bak-Geller Corona, looks at the construction of a nationalist gastronomy in the period following New Spain's independence from Spain. She demonstrates how the

European discourse on *bon gout* (good taste) was appropriated within Mexican nationalist discourse and ideology. Examining the parallels and divergences between French hegemonic and Mexican nationalist culinary practices she shows how modernity contributed to discursively shape the taste of Mexican gastronomy. Chapter 3 by Héctor Hernández Álvarez and Guadalupe Cámara Gutiérrez, "Alcohol Consumption Patterns among Different Social Groups during Yucatán's Gilded Age," examines, through the analysis of glass shards and other remains found in the refuse of nineteenth-century henequen plantations, the differences in practices of alcohol consumption between owners, their foremen, and the indentured workers on their properties. They show that there were different channels for the distribution of locally distilled alcohol and imported fine liquors, beers, and wines. Likewise, the differences in what was consumed and the social practices of consumption demonstrate the mechanism that articulates processes of social inclusion and exclusion among different social groups in the henequen plantations of Yucatán. Chapter 4 by Mario Fernández-Zarza and Ignacio López-Moreno, "The Flavors of Corn: A Unique Combination of Tradition and Nature," examines the importance of corn in the diet of people of the Mexican highlands. They affirm the quality of corn as a "superfood" and describe the benefits indigenous people draw from its consumption. As their chapter shows, corn is a culinary ingredient that has many different potential flavors, some derived from their traits at the moment of harvest (young to ripe), some from the production technique (boiled, roasted, ground, whole), and some that they draw from the other ingredients that accompany it. They argue for the need to foster the diversity of this grain against the industrial mass production of corn-derived products.

Part Two, "The Identity and Politics of Mexican Foods—and the Politics of Identity," begins with Chapter 5, by Ronda L. Brulotte, "A Taste for Agave: The Emerging Practices and Politics of Mezcal Connoisseurship." In this chapter, Brulotte argues that interdiscursivity plays an important part in shaping the linguistic code consumers and foodies use to describe mezcals at a time in which this distilled drink is transitioning from marking peasant taste to that of the upper classes with larger cultural and symbolic capital. As she demonstrates, consumers transfer descriptive codes from different edible and drinkable commodities, but also deploy the positive values associated with small batch and craft production, thus shaping the connoisseurship of mezcal drinkers. In Chapter 6, "Making and Changing Yucatecan Taste in Yucatán: Innovation and Persistence in Yucatecan Gastronomy," Steffan I. Ayora-Diaz looks at the effects of both the inclusion of "Mexican Food" in UNESCO's list of "Humanity's Intangible Cultural Heritage" on the discourse on Yucatecan food and also the changes in taste brought about by the explosion of the regional foodscape with a multiplication of efforts to innovate within regional cuisine, whilst others try to affirm and preserve a cooking style that some perceive as endangered. Thus, contemporary changes in the regional-national foodscape are contributing to regional changes in the importance of food in shaping the politics of Yucatecan cultural identity where age, place of origin, declared ethnicity, and nationality multiply individual dispositions toward different culinary traditions. In Chapter 7, "The Life Delicious: Taste and Politics in Mérida, Yucatán," Gabriela Vargas-Cetina examines how under the common umbrella of a positive understanding of life as "delicious," there are class-related differences in taste. These differences are expressed at different times in the local annual calendar when people of Mérida enjoy life through food, drinks, parties, and different forms of informally and formally organized sociality. As she illustrates through the description of annual celebrations and festivities, people of all social classes get together, albeit in different spaces, and consume foods and drinks of different qualities. However, it is through the shared meaning of celebratory times that they come to develop a common understanding of Yucatecan's delicious life, which masks differences in class and taste. Chapter 8, written by Jeffrey H. Cohen and Paulette K. Schuster, "To Eat *Chapulines* in Oaxaca, Mexico: One Food, Many Flavors," shows how the fashionable roasted grasshopper has different meanings for different groups of people within and outside the state of Oaxaca, Mexico. As the Food and Agriculture Organization (FAO) and nutritionists all over the world are singing the praises of edible insects, Cohen and Schuster show that for the Oaxaca peasants who harvest and prepare them for consumption, chapulines are a necessity and not a meal that would trigger nostalgia. This is different, they demonstrate, for the middle and upper classes in the state and Mexico who take the consumption of chapulines to be an indicator of indigenous heritage and culture, and seek to integrate it within the menus of touristic restaurants of the Oaxaca cities. Also, for tourists, chapulines are consumed in a quest for authentic, "exotic" foods and is also an edible item to dare their friends to eat.

Part Three, "Taste and Displacement: Mexican Food in the World of Consumption," begins with Chapter 9 by Ramona L. Pérez. In her chapter, "The Taste of Oaxaca: It's to Die for!" she also takes UNESCO's initiative regarding Mexican cuisine to explain how the media have exploited the indigenous origins of Oaxaca to advance it as a touristic place of choice: authenticity and tradition affirm the specific taste of the food of Oaxaca. She describes the heterogeneous landscape of Oaxaca and thus the local differences in privileged cooking ingredients. She points out how the lead-rich clay casseroles and other cooking instruments impart a flavor desired and sought after by people from Oaxaca, showing that in cooking taste often comes before health considerations. As she demonstrates, UNESCO's recognition of indigenous traditional cooking has had the unintended consequence of fostering the use of lead-rich cooking utensils. In Chapter 10, "Dos Equis and Five Rabbit: Beer and Taste in Greater Mexico," Jeffrey M. Pilcher looks at the transition from pulque to beer in Mexico and the creation of "Mexican" microbrewery beers in the United States. He discusses how since the nineteenth century pulque, the drink of choice in central Mexico while wine was scarce, progressively became accepted by all social classes. However, the arrival of beer and breweries influenced by German and US technologies put the prize on hygiene and modernity, displacing pulque and eventually marking it a lower-class drink, while beer became the preferred drink of the middle classes. The taste of Mexican beer, he argues, became important both for Mexican immigrants into the United States, and for US tourists who returned from trips to Mexico. This has allowed for a small brewing industry seeking ways to impart "Mexican" taste onto beers brewed in the United States. In Chapter 11, "Diffused Palates: The Evolution of Culinary Tastes of Jewish Mexicans Living in Israel," Paulette K. Schuster looks first at Jewish immigration in Mexico and discusses the culinary rules and preferences of the different groups that arrived from different European regions. She discusses the adaptation of Jewish dishes to Mexican ingredients and the adoption of Mexican recipes as part of the Jewish–Mexican taste. However, the Israeli state has attracted descendants from these different groups back to Israel, and in their travels those people of Jewish ancestors but with Mexican taste have had to negotiate the meaning and taste of Mexican food in Israel. In Chapter 12, "Defining Sanitized Taste and Culinary Tourism in Cozumel, Mexico," Christine Vassallo-Oby examines how food is defined and perceived by tourists coming to Mexican shores in the contemporary political arena in which the US president condemns Mexicans in different negative ways. As she discusses, the cruise companies strongly suggest that eating in places not controlled by them may be dangerous for travelers. They implicitly dwell on images of Mexicans and Mexico as dirty and dangerous. Hence, tourists are encouraged to eat at the local outposts of US franchises, consuming foods they are familiar with at home, or to experience versions of Mexican food preapproved by the cruise ship company. Finally, in the Postface, Richard Wilk highlights the paradoxes that surround the imagination of national cuisines in a globalized economy in which recognitions and certifications, such as that by UNESCO, can be turned into an added value for national, regional, and ethnic cuisines, culinary practices, and ingredients. These paradoxes become salient during the negotiation and establishment of what a national cuisine is, both for those at "home" and those abroad, be it migrants, those who choose between homogenized cuisines based on cultural stereotypes, and those who choose their food for its diversity.

In different ways, the chapters of this volume show different levels and scales of identity politics in which Mexican food is either affirmed or challenged, both from within and from without. Food and drinks each have a particular history that has unfolded under changing economic, political, and ecological circumstances. Tourism, migration, and immigration, the displacement of members of different ethnic groups, religion, and discourses on local, regional, ethnic, and national identities, all contribute to the reproduction of ambivalent and sometimes conflictive representations in which food and drinks play an important part as mediators and carriers of politically charged meaning.

Note

1 The field of food and nutrition studies is slowly growing in Mexico, as is scholarship about Mexican food; still, there is scant attention in Mexican academia to social, cultural, and political issues surrounding food taste and aesthetics, and their relationship with political affirmations of identity—with the exception of nationalism and food. Fortunately, interest is now growing, and hopefully we will soon find more studies on this and related issues by researchers based in Mexico. It is one of our hopes that this volume will contribute to trigger and encourage more studies into taste and the different sensorial dimensions of food.

References

Agamben, G. ([1979] 2017), *Taste*, trans. C. Francis, London: Seagull Books.

Appadurai, A. (1981), "Gastro-politics in Hindu South Asia," *American Ethnologist*, 8 (3): 494–511.

Appadurai, A. (1988), "How to Make a National Cuisine: Cookbooks in Contemporary India," *Comparative Studies in Society and History*, 30 (1): 3–24.

Ayora-Diaz, S. I. (2010), "Regionalism and the Institution of the Yucatecan Gastronomic Field," *Food, Culture and Society*, 13 (3): 397–420.

Ayora-Diaz, S. I. (2012a), *Foodscapes, Foodfields and Identities in Yucatán*, Amsterdam: CEDLA, New York: Berghahn.

Ayora-Diaz, S. I. (2012b), "Gastronomic Inventions and the Aesthetics of Regional Food: The Naturalization of Yucatecan Taste," *Etnofoor*, 24 (2): 57–76.

Bak-Geller Corona, S. (2013), "Narrativas deleitosas de la nación. Los primeros libros de cocina en México (1830–1890)," *Desacatos*, 43: 31–44.

Bhabha, H. (1994), *The Location of Culture*, London: Routledge.

Bourdieu, P. ([1979] 1984), *Distinction: A Social Critique of the Judgment of Taste*, trans. R. Nice, Cambridge, MA: Harvard University Press.

Bourdieu, P. (1991), *Language and Symbolic Power*, Cambridge, MA: Harvard University Press.

Braidotti, R. (1994), *Nomadic Subjects: Embodiment and Sexual Difference in Contemporary Feminist Theory*, New York: Columbia University Press.

Camporesi, P. (1970), "Introduzione," in P. Artusi, ed., *La scienza in cucina e l'arte di mangiare bene*, vii–lxxxi, Turin: Giulio Einaudi Editore.

Capatti, A. and M. Montanari (1999), *La cucina italiana. Storia di una cultura*, Bari: Laterza.

Cwiertka, K. J. (2006), *Modern Japanese Cuisine: Food, Power and National Identity*, London: Reaktion Books.

Descola, P. ([2005] 2013), *Beyond Nature and Culture*, trans. J. Lloyd, Chicago: University of Chicago Press.

Edensor, T. (2012), *National Identity, Popular Culture and Everyday Life*, Oxford: Berghahn.

Ferguson, P. P. (2004), *Accounting for Taste*, Chicago: University of Chicago Press.

Finn, S. M. (2017), *Discriminating Taste: How Class Anxiety Created the American Food Revolution*, New Brunswick, NJ: Rutgers University Press.

Foster, M. D. (2015), "UNESCO on the Ground," in M. D. Foster and L. Gilman, eds., *UNESCO on the Ground: Local Perspectives on Intangible Cultural Heritage*, Bloomington: Indiana University Press, Kindle Edition.

Gaytán, M. S. (2017), "The Transformation of Tequila: From Hangover to Highbrow," *Journal of Consumer Culture*, 17 (1): 62–84.

Gronow, J. (1997), *The Sociology of Taste*, London: Routledge.

Hechter, M. (1975), *Internal Colonialism: The Celtic Fringe in British National Development*, Berkeley: University of California Press.

Holmes, B. (2017), *Flavor: The Science of Our Most Neglected Sense*, New York: W. W. Norton.

Howes, D., ed. (2005), *Empire of the Senses: The Sensual Culture Reader*, Oxford: Berg.

Irvine, J. T. and S. Gal (2003), "Language Ideology and Linguistic Differentiation," in P. V. Kroskrity, ed., *Regimes of Language: Ideologies, Polities, and Identities*, 35–83, Santa Fe, NM: New School of American Research.

Johnston, J. and S. Baumann (2010), *Foodies: Democracy and Distinction in the Gourmet Foodscape*, London: Routledge.

Korsmeyer, C. (1999), *Taste: Food and Philosophy*, Ithaca, NY: Cornell University Press.

Korsmeyer, C., ed. (2017), *The Taste Culture Reader: Experiencing Food and Drink*, 2nd edn., London: Bloomsbury.

Lane, C. (2014), *The Cultivation of Taste: Chefs and the Organization of Fine Dining*, Oxford: Oxford University Press.

Latour, B. (1983), *We Have Never Been Modern*, Cambridge, MA: Harvard University Press.

Long, J. (1992), "The Mexican Contribution to the Mediterranean World," *Diogenes*, 40 (159): 37–49.

Magaña González, C. R. (2016), "Technologies and Techniques in Rural Oaxaca's Zapotec Kitchens," in S. I. Ayora-Diaz, ed., *Cooking Technology: Transformations in Culinary Practice in Mexico and Latin America*, 55–67, London: Bloomsbury Academic.

McQuaid, J. (2015), *Tasty: The Art and Science of What We Eat*, New York: Scribner.

Moncusí, A. and B. Santamarina (2008), "Bueno para comer, bueno para patrimonializar. La propuesta de la cocina Mexicana como patrimonio inmaterial de la humanidad," in M. Álvarez and F. X. Medina, eds., *Identidades en el plato. El patrimonio cultural alimentario entre Europa y América*, 127–142, Barcelona: Icaria.

Mouritsen, O. G. and K. Styrbæk (2017), *Mouthfeel: How Texture Makes Taste*, New York: Columbia University Press.

National Institute of Geography and Statistics (INEGI) (2015), "Lenguas indígenas en México y hablantes (de 3 años y más) al 2015," INEGI, http://cuentame.inegi.org.mx/hipertexto/todas_lenguas.htm (accessed September 20, 2018).

Pilcher, J. M. (1998), *Que vivan los tamales! Food and the Making of Mexican Identity*, Albuquerque: University of New Mexico Press.

Pilcher, J. M. (2006), *The Sausage Rebellion: Public Health, Private Enterprise, and Meat in Mexico City, 1890–1917*, Albuquerque: University of New Mexico Press.

Pilcher, J. M. (2012), *Planet Taco: A Global History of Mexican Food*, Oxford: Oxford University Press.

Senado de la República Mexicana (2016), "Iniciativa con proyecto de decreto por el que se expide la Ley Federal de Fomento a la Gastronomía Mexicana," LXIII Legislatura, www.senado.gob.mx/sgsp/gaceta/63/1/2016-04-28-1/assets/documentos/Dic_Ley_Fomento_Gastronomia.pdf. (accessed April 14, 2018).

Shepherd, G. M. (2012), *Neurogastronomy: How the Brain Creates Flavor and Why It Matters*, New York: Columbia University Press.

Spence, C. (2017), *Gastrophysics: The New Science of Eating*, New York: Viking.

Stillman, J. A. (2002), "Gustation: Intersensory Experience par excellence," *Perception*, 31 (12): 1491–1500.

Stoller, P. and C. Olkes (1989), "The Taste of Ethnographic Things," in P. Stoller, ed., *The Taste of Ethnographic Things: The Senses in Anthropology*, 15–34, Philadelphia: University of Pennsylvania Press.

Strong, J., ed. (2011), *Educated Tastes: Food, Drink and Connoisseur Culture*, Lincoln: University of Nebraska Press.

Stuckey, B. (2012), *Taste: What You're Missing*, New York: Free Press.

Sutton, D. E. (2001), *Remembrance of Repasts: An Anthropology of Food and Memory*, Oxford: Berg.

20Minutos (2018), "Certificarán a cocineras tradicionales en México," February 23, www.20minutos.com.mx/noticia/336511/0/certificaran-a-cocineras-tradicionales-en-mexico/ (accessed April 2, 2018).

United Nations Educational, Scientific and Cultural Organization (UNESCO) (2010), "Traditional Mexican Cuisine: Authentic, Ancestral, Ongoing Community Culture, the Michoacán Paradigm" [Documentary], Conservatorio de la Cultura Gastronómica Mexicana, www.unesco.org/archives/multimedia/?pg=33&s=films_details&id=1674 (accessed September 20, 2018).

Wilk, R. (2010), "Power at the Table: Food Fights and Happy Meals," *Critical Studies Critical Methodologies*, 10 (6): 428–436.

Wilson, A. (2013), *The Industrial Diet: The Degradation of Food and the Struggle for Healthy Eating*, New York: New York University Press.

Part One

Tasting the Past in Mexican Foods

1

A Touch of Pre-Columbian Maya Flavor

Lilia Fernández Souza,
Universidad Autónoma de Yucatán

Archaeology, food, politics, and the senses

Archaeology is an anthropological discipline that studies the past through material evidence. Through archaeological evidence we can debate subsistence, economic, religious, and political issues, to mention a few of the possible topics of interest about the human past. There are questions yet to be fully answered: can we understand the senses and feelings of ancient people? How close can we get to what people felt, tasted, smelled, enjoyed or chose to eat and drink? Was there a relation between taste and politics? Are these good questions? Does it even matter?

Food has always been intimately involved with politics. Hastorf (2017) affirms that the presentation of food is a way to display differences of power as well as ideological asymmetries. Food may be used to stress inequality, exclusion, and differentiated access to certain products but, on the other hand, banquets and feasts may be occasions to share and negotiate or to create and to reinforce community ties. According to Hastorf (2017: 182), "Class is one of the most common delimiters of culinary rules." Manners, etiquette, sequence of service, and presentation are some of the ways of marking social, economic, and political differences. Trends are imposed in different ways. For example, Bourdieu (1984) reflects about the construction of "legitimate cultures" that indicate the best-accepted trends to perform practices such as the appreciation of arts and the tastes of food. Individual choices depend both on family social background and on formal education that provide the blueprint about what is considered elegant, cultivated, and culturally legitimate. In certain regards the social origins of individuals are a heavier factor than education, but in any case "This predisposes tastes to function as markers of 'class'" (Bourdieu 1984: 2).

Hence, complex societies have different sorts of cooking practices and eat according to socio-economic and political differences. Stephen Mennell (2005), based on his study of medieval Europe, discusses the mechanisms where *haute cuisines* are created. He argues that social stratification is necessary but not sufficient for the emergence of an elaborate culinary taste, adding another required ingredient: long chains of social interdependence. Mennell (2005: 26, 27) argues that when there exists a deep social division and an unequal interdependence between strata, the power and status of the elites is manifested through the quantity of food, not its quality nor the complexity in food preparation. But when social strata get closer and interdependence becomes more equal, there exists more competition and cuisines change to become more elaborate. On the other hand, long interdependence chains also allow access to a wider variety of ingredients. Mennell (2005: 27) defines haute cuisine "by their typical dishes requiring complex sequences of stages and considerable division of labour among kitchen staff," and suggests that it first emerged in court societies such as that of ancient Egypt. Hence politics impacts culinary practices in several ways: different social strata have distinct access to alimentary products, and the elites find strategies to stress their membership to a legitimate culinary culture and, consequently, to refined and sophisticated tastes of haute cuisine.

Different studies have stressed the fact that in Western tradition taste has been devaluated in favor of vision and hearing (Cárdenas Carrión 2014; Hamilakis 2011: 209; Sutton 2010: 210). Nonetheless, there is a growing number of anthropologists interested in how the senses, and particularly taste, are a fundamental part of human experience. Gustemology, as defined by Sutton (2010: 215), refers to "such approaches that organize their understanding of a wide spectrum of cultural issues around taste and other sensory aspects of food." Such issues include how and why people decide that something is edible or inedible (Smith 2006: 480); how taste practices define what is appropriate or inappropriate, and become part of the formation of the self and of a group (Hastorf 2017); how food plays an important role in identity processes (Ayora-Diaz 2012); and how good and bad taste intervene in social interaction and negotiation (Stoller and Olkes 1989).

Often, taste is reduced to a chemical reaction of the taste buds, but we must recognize that it is also a multisensorial experience (Sutton 2010: 211). The human sensorium is based on synesthesia, the interactive union of all the senses (Cárdenas Carrión 2014: 35; Hamilakis 2011: 210; 2015; Sutton 2010: 218). As Hamilakis (2015: 205) points out, eating involves more senses than taste and smell: it uses tactility, visual tactility, the sense of emplacement, memory, and intoxication. Food substances, human bodies, artifacts, things, places, and performances, all together create a particular sensorial assemblage (Hamilakis 2015: 205). In their influential work *The Taste of Ethnographic Things*, Stoller and Olkes (1989: 32) advocate for what they call *tasteful fieldwork*; meaning that the anthropologists should not only investigate kinship, exchange, or symbolism but also "describe with literary vividness the smell, tastes, and textures of the land, the people, and the food." This makes sense when we consider that everyday human life is full of a variety of sensual experiences which give meaning to daily activities, choices, and desires, and are imbedded with memory and emotion. So, we should produce *tasteful ethnography*. However, if what we want is to study ancient cultures, how possible is it to produce *tasteful archaeology*?

Archaeology has frequently raised questions about food in past societies: were they producers or appropriators? Were they hunters, pastoralists, fishermen, or farmers? Was their diet based on vegetables or meat? Were they egalitarian or hierarchical? Did people of a region take the same dietary decisions as their neighbors? Do dead people still need to eat? Answering these questions will allow us to understand some aspects related to sustenance, social and political organization, gender, world vision, religion, and identity among past cultures. However, as noted by different scholars (Hamilakis 2011: 208; 2015: 205; Hastorf 2017; Smith 2006: 481), archaeological research often focuses on aspects such as diet, production, subsistence systems, economy, logistics, and technology to make food digestible, but rarely on sensorial experiences of eating.

Sensory experiences seem to be ephemeral and intangible, and they may seem the opposite to archaeological evidence. Nonetheless, Hamilakis (2011: 209) claims that "sensory experience is material, it requires materiality in order to be activated, and its past and present material traces are all around us, whether it is the burnt bones of a pig that was sacrificed and then consumed, or the traces left on a rock which was repeatedly hit deliberately to produce sound." The study of senses in archaeology does not pretend to bring the past sensorial experience to our present, or to make us feel the same taste or emotion that someone else felt in the past: that is an impossible quest, as sensual experiences are culturally and historically situated (Classen and Howes 1996: 87; Comis and Corrado 2009; Hamilakis 2011: 208). Archaeology can analyze sources and material remains that evoke flavors, textures, and aromas. Such is the case of the experimental recreation of Roman *garum* following both archaeological and historical evidence (Comis and Corrado 2009). As Hastorf suggests (2017: 392–393): "The materiality of food allows us to complete an archaeology of human experience, as both material and emotional aspects of a lived life surround food."

In this chapter, I offer an approach to Maya pre-Columbian flavors. I discuss the evidence that may bring us close to the taste and aromas on the ancient Maya land, and to the differences between commoners' and elites' consumption. To accomplish these goals, archaeology must turn to multiple additional sources, such as botany and ecology (Matos Llanes and Acosta Ochoa 2016; Zimmermann 2008; Zizumbo Villarreal, Flores Silva and Colunga García-Marín 2012; Zizumbo Villarreal, García and Colunga García-Marín 2008); zooarchaeology (Götz 2014); chemical analysis of food residues (Barba, Ortiz and Pecci 2014; Henderson et al. 2007); and epigraphy and iconography to decipher writing and interpret images (Beliaev et al. 2009; García Barrios 2017; García Barrios and Carrasco Vargas 2008; Kantun-Rivera 2010; Kettunen and Helmke 2011; Stuart 2006; Vail 2009). Finally, we must also take into account historic texts, dictionaries, ethnographic records, and the contributions of ethnoarchaeology to complete a multidisciplinary approach (Arzápalo Marín 2003; Barrera Vázquez 1995; Garza et al. 2008).

Ingredients and their history

The most prominent ingredients of the Mesoamerican pre-Columbian diet are the well-known triad of maize, beans, and squash. Far from a simple and monotonous menu, research has unveiled a very complex scenario. These cultigens were neither domesticated at the same time nor in the same place. There is evidence of domesticated squash between 8900 and 7900 BP (Before Present), of corn around 9000 BP, and of common beans (*Phaseolus Vulgaris*) around 8000 and 6000 BP. Domesticated chili peppers (*Capsicum Anuum*) have been found at about 6000 BP, and a variety of tubers around the fourth millennia BP (Pohl, Pope and Jones 2000: 259; Zizumbo Villarreal, García and Colunga García-Marín 2008: 93; Zizumbo Villarreal, Flores Silva and Colunga García-Marín 2012: 330).

Corn was the undisputable king of the Mesoamerican diet. According to Clark et al. (2007: 23) and Rosenwig (2006), the importance of maize grew from 1000 BCE when, for some reason, Mesoamerican people chose corn over other cultigens such as tubers (like *makal* [*Xantosoma yucatanense*] and sweet potatoes), without excluding the latter from their diet. For the pre-Columbian Maya, archaeological and historical data suggest mainly a vegetarian diet, enriched with game or fish, especially for elite groups in wealthier sites (Garza et al. 2008; Götz 2014). Apparently, at least in some contexts, meat was quite appreciated: the Dresden Codex, a postclassic northern Lowland Maya document, has, among other topics, a description of New Year ceremonies in which food was offered to major gods and goddesses (Vail 2009). Meals and beverages of corn and cacao were ubiquitous, as well as meat dishes; there were tamales stuffed with venison, fish, iguana, turkey and other birds, and also platters of meat and jars full of alcoholic beverages. Early colonial sources, such as Bishop Diego de Landa's *Relación de las cosas de Yucatán*, describe similar ceremonies in which the principals of the towns and common people ate delicacies like quails and venison's hearts. According to Farris (2012: 240), festivities were important occasions, both in religious and social senses. They were the moments in which commoners had access to foods and drinks that were not part of their daily life. After the conquest, pork lard became an especially appreciated delicacy (Farris 2012).

Early colonial sources offer information about the amazing variety of vegetal and animal resources that were consumed by the Yucatec Maya (Garza et al. 2008; Landa 2001; Roys 1972): in addition to corn and different types of beans, squashes, and chili peppers, there were fruits, leaves, seeds, roots, and tubers, as well as mammals, birds, and fishes from the sea and sinkholes (*cenotes*). There were few domesticated animals—dogs, turkeys, and bees—so most of the meat came from hunting and fishing. Native bees, particularly the *Melipona becheii*, a stingless bee, produced a priceless honey, which was used both as a sweetener and as the base for most alcoholic beverages. Yucatán was and still is rich in salt from the peninsular coastal salt mines. After the conquest, new animals, vegetables, and other ingredients were imported, although their influence and adoption depended on such factors as the distance from settlements to cities and the limited purchase power of the inhabitants of towns. As argued by Ayora-Diaz (2012, 2017), today Maya food is only one of multiple components of the contemporary Yucatecan kitchen. Some culinary practices can be traced back centuries in time, and several current cooking practices are helpful to understand material evidence from the past.

Flavors, blends, and textures

From around 1000 BCE, corn was one of the pillars of Mesoamerican and Maya meals. Pre-Columbian Maya sources suggest that the *tamale* was the most common form of preparation. In an influential work, Taube (1989) examined the iconography and texts from Maya pottery and sculpture and found different sorts of tamales: some of them were rounded, with a sort of notch on their top that suggests that they were stuffed. Others looked like rolls, and others were clearly covered with some dripping substance, like a sauce. As mentioned above, in Maya art and texts there are clear examples of meat tamales. In Maya languages, the word for tamale is *waaj*, which is also used for "tortilla" and other corn meals. In early Yucatecan colonial dictionaries, a translation for *waaj* was *pan*, "bread" in Spanish. Some of the mentioned dishes were *bakil waaj* (meat patty), *kayil waaj* (fish patty), *hak sikil* (beans and squash seed bread), *thancabal waaj* and *pim waaj* (thick tortillas), and *liktzuhil waaj* (fresh bread) (Arzápalo Marín 2003: 252). Bishop Landa (2001: 75) describes some other varieties, which included ingredients such as quail meat, egg yolk, venison heart, and "their dissolved pepper" (chili pepper).

Today, tamales are still appreciated in Yucatán, and they are a frequent dish for birthdays and other parties. It is relatively common to find successful street tamale vendors in some corners of Mérida city, because preparation at home is relatively laborious. Tamales or tamale-like dishes are diverse in form, size, flavor, and texture: some are steam cooked, such as *vaporcitos, tamales colados* (strained tamales) and *dzotobichayes* (rolled chaya leaf tamale), while others like *mucbil pollo* (a special dish for the Day of the Dead celebrations) and *chachak waajes* are baked in an underground oven named *pib*. In all cases, the base is corn dough, stuffed, and covered or wrapped with maize or banana leaves before cooking. But, depending on the sort of tamale, the dough can be mixed with *xpelon* beans, chaya leaves, or achiote seeds (*Bixa orellana*) milled and dissolved in water. The stuffing depends on access to ingredients and individual preference, and it may be made out of lima beans, poultry, or pork. *Dzotobichay* is made without meat: it is a long roll of corn dough mixed with finely chopped chaya leaves and stuffed with roasted ground squash seeds. The variant called *brazo de reina* ("queen's arm") includes boiled eggs, and both are served covered with tomato sauce and milled roasted squash seeds. The latter tamale is especially appreciated during Lent.

Tamales offer a wide variety of textures: strained tamales are very delicate and soft, while *mucbil pollos* are (or may be) crunchy, and some *polkanes* ("snake heads," small ellipsoid tamales stuffed with lima beans and squash seeds) are so hard that some people call them *piedras* (stones). There are a few differences between contemporary and pre-Columbian tamales, for it was the Spaniards who brought poultry, pork, and beef, as well as banana leaves. But perhaps the most relevant difference is the use of lard in the mixture of the dough, which did not exist in pre-Hispanic times. For many contemporary Yucatecans, the flavor and texture that lard gives to corn-made dough, fried beans, and other dishes is irreplaceable. Chef Wilson Alonso, who is a professor and researcher of ancient Maya cuisine at the Technologic University of Maxcanú, in Yucatán, cooked *mucbil pollos* without lard achieving a less greasy and unctuous, but still delicious, result (Alonso, pers. comm., November 2017).

Another element in tamales and other dishes is *k'ol*, a thick sauce/gravy made with maize dough dissolved in water or fowl broth, mixed with salt and achiote seeds. *K'ol* is an important part of the stuffing for some tamales but also may be a meal in itself. During some peasants' ritual celebrations, a large container of *k'ol* is prepared with chicken broth and achiote, whipped vigorously until cooked, and served in smaller bowls to be collectively eaten. The *k'ol* is left to cool for a while, forming a curd, while the religious celebrant or *hmen* prays, and later it is taken out from the container and cut into pieces for serving. The final product is somewhat more solid than jelly or flan.

Corn was the main ingredient in many pre-Columbian beverages, including a wide variety of combinations, both sweet and salted. The Primary Standard Sequence (PSS) or Dedicatory Text is a written formula found in the rim of elite ceramic containers which describes the kind of pottery piece (*ulak* [his or her dish], *ujay* [his or her bowl], *yuk'ib* [his or her drinking vase], *ujawante'* [his or her legged dish]), its decorative treatment (painted or carved), its content, and, sometimes, the name of its owner. From the PSS epigraphers have found mentions of maize *atole* beverages such as *'ul, sa'* and *sak ha* and variants such as *ch'aj ul* ("bitter atole"), *kakawal ul* (atole with cocoa), *pah ul* (sour atole), and *is ul* (atole with sweet potatoes) (Beliaev et al. 2009). *Sak'ul*, or white atole, was a beverage prepared with tender corn and was deposited in the graves of the deceased (García Barrios 2017). Much less elaborated but ancient (probably of pre-Columbian origin) is *pozole* or *keyem*, ground nixtamalized (soaked in water with lime) corn kneaded to form balls that people take to work for lunch. When people are hungry, they dissolve the *keyem* in water and drink it. Today, a very popular way to enjoy *keyem* is to have a sip of it and immediately bite a hot chili pepper with a bit of salt. *Keyem* may be also sweetened with honey or sugar; and many people like "pozole con coco," *keyem* with grated coconut. Atole (*ul* or *sa*) and *pozole* or *keyem* have different textures: the first one is smooth, while *keyem* has chunks that separate from the liquid and sink to the bottom of the cup. This quality induces a characteristic body movement, as the hand is swirled in circles (slightly moving the arm) to blend the sediment with the liquid. Hence, *keyem* is a beverage that also contains something edible that people find satisfying.

Without any doubt, chocolate was one of the most appreciated beverages in Maya pre-Columbian times. Chemical residue analysis has offered evidence of theobromine, cocoa's chemical fingerprint, in vessels from a number of sites such as Rio Azul, Guatemala, and Puerto Escondido, Honduras (Henderson et al. 2007). There exists extensive epigraphic evidence of many ways for enjoying chocolate: Stuart (2006: 193) mentions, for example, *tzih kakaw* (pure cacao), *ach' kakaw* (new cacao), and *yutal k'an kakaw* (ripe cacao). Kettunen and Helmke (2011: 45) describe *chak kakaw* (red cacao) and *om? kakaw* (frothy cacao). Beliaev et al. (2009) mention *yutal kakaw* (fruity cacao; maybe cacao with fruits or a beverage prepared with cocoa fruit flesh instead of the seeds), *kaab'il* or *chab'il kakaw* (cacao with honey), and *tzah kakaw* (sweet cacao). Some epigraphic readings are still obscure. For example, Beliaev et al. (2009) report a word with probable termination in *-tz*. After study, they suggest it may be the Maya Chorti word *sutz*, meaning, "any fruit like a berry," and consider the *capulin*, a cherry-like small fruit, as a possibility. According to paleobotanic research conducted by Lentz (1999: 12), *capulin* was one of the fruits of which exists evidence in one or more pre-Columbian Maya sites. In another interesting case, Tokovinine (2016) deciphered a text in a vase from Rio Azul, Guatemala, and proposed the reading as *y-uk'ib ta ibil kakaw*: "his vase to drink *ib* cacao"; *ib* is the name of lima beans (*p. lunatus*), which may offer another cocoa flavor. Ingredients of other recipes have been found in archaeological contexts. For example, Matos Llanes and Acosta Ochoa (2016) reported starch grains of cocoa, corn, and achiote from a grinding stone at the Classic site of Sihó. While it is difficult to guarantee that they were grinded at the same time, it is interesting to note that there still exists a beverage made of those ingredients called *tascalate*, which is not common in Yucatán but in Chiapas, a southern Mexican state. In pre-Columbian times, as previously mentioned, chocolate was often mixed with atole. Beliaev et al. (2009: 265) describe beverages such as *kakawal 'ul*, atole with a bit of cocoa, which was different to *sa'al kakaw*, cocoa beverage with a liquid dough consistency or with some atole. Finally, it is important to stress the relevance of cocoa's provenience: Kettunen and Helmke (2011: 45) suggest that the PSS sometimes offers toponyms referring to the chocolate in the vessel, acting like a certificate of origin or, at least, demonstrating that people paid attention to where the cocoa they were drinking came from.

In addition to ingredients and combination of flavors, there were also several techniques of preparation. Grinding was one of them, which left a ubiquitous and permanent artifact: the grinding stone, *metate* or *k'a* (Fernández Souza 2016; Fernández Souza, Toscano and Zimmermann 2014). Chemical analysis of soil around big, legless *metates* in the Classic sites of Kabah and Sihó showed distinct concentration patterns of residues including carbonates and protein (Fernández Souza, Toscano and Zimmermann 2014; Matos Llanes and Acosta Ochoa 2016), suggesting the possibility that some grinding stones were used to grind specific foods, or that there was a deliberate attempt not to mix flavors. There also exists data of a wide diversity of cooking techniques, both in archaeological and historic sources. Götz (2014: 184) analyzed zooarchaeological bones of animal, such as white-tailed deer, and reported evidence of boiling and dried-bone fractures, few cut marks, as well as a scarcity of evidence of direct fire. He believed that this might be the result of baking in an underground oven, *o pib,* which can reach temperatures of 200–300 degrees Celsius without direct fire on meat or bones. Götz (2014: 186) also reported evidence of roasting and smoking, with a few examples of direct fire, especially in coastal sites. All these techniques were reported in early colonial dictionaries too (Barrera Vázquez 1995). There were also meals prepared with broth or other liquids. In Sihó, a Classic Maya site, spot test chemical analysis in fragments of pottery jars and basins demonstrated the presence of differentiated levels of residues of phosphates, carbonates, fatty acids, carbohydrates, and protein residues. Jars are narrow-necked pieces, so chemical results suggest they contained liquids with organic substances, and differentiated levels of sugars, lime (probably nixtamalized corn), and proteins. Fragments of pottery from palace-type structures showed higher levels of protein residue than nonelite structures, suggesting more meat consumption. In such cases, jars with elevated protein residues could have been used to prepare meat stews.

Getting closer to ancestral taste

Until now, no written pre-Columbian Maya recipes have been found: we don't know the quantities they used nor the cooking time for meals. Being realistic, it is likely that we will not have them in the near future—or maybe ever. Nevertheless, the conjunctive approach, combining archaeological and anthropological methods and concepts, allows us to put together several clues and provides us with many hypotheses to test and try through experimental archaeology. Evidence teaches us many things that give a colorful idea of how rich and diverse the foodscape was in pre-Columbian Maya times. In this chapter I have provided a general overview and suggested that there were differences between geographic and political regions, sites, socio-economic stratification and historical moments. As mentioned above, we cannot bring back the experience of people of the past when they tried flavors and smells. But there are a number of facts we can examine and discuss. First, Maya pre-Columbian diet was wide and varied, and included uncountable vegetables. Corn was the most important crop followed by beans, tubers such as sweet potatoes and jicamas, chili peppers, squash, cocoa, chaya leaves, and fruits like mamey, plum, capulin, avocado, sapote, and dragon fruit. Many of these had varieties that were used and consumed in different forms and combinations. Although evidence suggests that most of the diet was vegetal, it also included game such as venison, rabbit, peccary, gopher, curassow, and manatee, and a variety of fish and mollusca from the sea and sinkholes, plus domesticated turkeys and dogs (Götz 2014).

Second, preparation and cooking techniques included grinding, straining, toasting, boiling, roasting, steaming, and baking. Historic and contemporary research also shows that some foods, such as chili peppers and eggs, were and still are cooked in the ashes of the hearth (Fernández Souza 2016). Some of the above-mentioned dishes and beverages require a combination of techniques, including nixtamalization and wrapping in leaves, and the same ingredient develops different flavors and textures depending on the elaboration processes. For example: to prepare chocolate, people roast, peel, and grind cocoa seeds, but the outcome varies according to how many times and how finely the seeds are ground. Finely ground cocoa seeds may be diluted in water to make a thin and light chocolate drink; when vigorously whipped, the liquid becomes frothy, an appreciated quality since pre-Columbian times (Reents-Budet 2006). In contrast, others like coarsely ground cocoa seeds mixed with some water and sugar (honey in pre-Hispanic times), as they enjoy chewing the chunks and the experience of simultaneous sweetness and bitterness on the tongue. Also, chocolate diluted in atole has a thick and silky liquid texture, with no chunks at all. Furthermore, as previously mentioned, the Maya combined cocoa with honey, fruits, achiote, probably lima beans, and even flowers. Henderson et al. (2007) have suggested the likelihood that cocoa fruit flesh was fermented to prepare an alcoholic beverage.

Third, the archaeological findings, as well as images and texts, strongly suggest that the Maya had the intention to obtain sweet, salty, bitter, spicy, and aromatic dishes and beverages. The use of seasonings such as achiote, chili peppers, salt, and honey is well documented. In addition to being the most important sweetener, honey was fermented with the bark of the balché tree (*Lonchocarpus longistylus*) and other roots to make alcoholic beverages (Chuchiak 2003; Garza et al. 2008). Some food and drink blends produce a combination of flavors (salty *keyem* with spicy chili pepper; the sweetness and alcoholic notes of honey;

the woody flavor of barks and roots; the bitterness of cocoa; and the fruity sweetness of capulin) and textures (the crunchy crust and the creamy stuffing of mucbil pollos; crispy roasted squash seeds with smooth and liquid atole; subtle froth floating on a silky chocolate beverage surface; and many others). Some aromas must have been quite similar to what we smell in contemporary Yucatecan dishes, such as when cooks use roasted chili peppers, boiled or baked ears of maize, boiled beans, baked fish with achiote, or thick tortillas toasted on the griddle. Some others are postconquest, especially those which incorporate flavors such as garlic, lime, onion, sugar, and, notably, pork lard.

And, fourth, there is the visual aesthetic experience of food. Achiote seeds, in addition to their flavor, provide a bright red color. Nowadays, achiote paste dissolved in water is used to color the maize dough and surface of some tamales; and *tascalate*, a chocolate drink with corn and achiote, has a particular reddish color. This suggests that some choices could have been (and still are) made for a visual reason and not only for taste and smell. On the other hand, as suggested by the texts of the SPP, food and drink were beautifully represented in elite pottery wares: a diversity of tamales, alcoholic beverages, and dripping chocolate vases were painted over or next to the thrones of the lords in their draped palaces. Those pieces were so delicately and exquisitely painted or carved they were like looking at the original. Although these contexts only express parts of the daily life of the classic Maya elites, such images open a door to a synesthetic atmosphere, where flavors, aromas, textures, colors, light, and sometimes music and dance, enveloped actions, social relations, and political decisions.

Final thoughts

One of the questions in this chapter, following Stoller and Olkes's question (1989), was: may we do *tasteful archaeology*? And, I would answer, *yes*. We may and we can. We definitely need to understand subsistence systems, technological improvements, and economic networks, but there is no reason for not asking those difficult questions about how human emotions and senses influenced decisions, choices, and social relations, and were strongly interrelated with politics and social complexity. In the pre-Columbian Maya world, food played a central role in public and private affairs, as viewed in classic Maya painted vessels, exclusively decorated for the elites. Beautiful wares and eating delicacies were part of political performances that expressed prominence through fine dishes, precious cacao beverages, and intoxicating drinks. The best menu was offered to the gods and goddesses: venison legs, platters with fishes, iguana tamales, chocolate, honey, and alcoholic drinks were served in public banquets controlled by the ruling families. In this way, the elite negotiated with divinities to keep the order of the universe, and commoners were part of that negotiation, participating in religious festivities with both devotion and desire for enjoying special food and drink to which they had no frequent access in their daily life. Also, material patterns of ingredients and cooking processes will contribute to our knowledge of culinary and gastronomic practices from the past, through analyzing the systematic and durable steps and procedures used that were expected to achieve specific flavors, textures, colors, and smells. As Hamilakis (2011) argues, sensory experience requires materiality to be activated. In the case of pre-Columbian Maya archaeology, multidisciplinary research is on its way to provide a solid material framework around the multisensorial experiences of the human past.

References

Arzápalo Marín, R. (2003), "Una perspectiva semiótica a la alimentación en Yucatán," in S. Poot Herrera, ed., *En gustos se comen géneros. Congreso Internacional de Comida y Literatura, Tomo I*, 343–368, Mérida: Instituto de Cultura de Yucatán.

Ayora-Díaz, S. I. (2012), *Foodscapes, Foodfields and Identities in Yucatán*, Amsterdam: CEDLA, New York: Berghahn.

Ayora-Diaz, S. I. (2017), "Translocalidad, globalización y regionalismo: cómo entender la gastronomía regional yucateca," *Anales de Antropología*, 51 (2): 96–105.

Barba, L., A. Ortiz and A. Pecci (2014), "Los residuos químicos. Indicadores arqueológicos para entender la producción, preparación, consumo y almacenamiento de alimentos," *Anales de Antropología*, 48 (3): 201–239.

Barrera Vázquez, A., ed. (1995), *Diccionario Maya*, Mexico City: Editorial Porrúa.

Beliaev, D., A. Davletshin and A. Tokovinine (2009), "Sweet Cacao and Sour Atole: Mixed Drinks on Classic Maya Ceramic Vases," in J. E. Staller and M. Carrasco, eds., *Precolumbian Foodways: Interdisciplinary Approaches to Food, Culture and Markets in Ancient Mesoamerica*, 227–257, New York: Springer.

Bourdieu, P. (1984), *Distinction: A Social Critique of the Judgement of Taste*, trans. R. Niece, Cambridge, MA: Harvard University Press.

Cárdenas Carrión, B. M. (2014), "Construcciones culturales del sabor. comida rarámuri," *Anales de Antropología*, 48 (1): 33–57.

Chuchiak, J. (2003), "'It Is Their Drinking that Hinders Them': Balché and the Use of Ritual Intoxicants among the Colonial Yucatec Maya, 1550–1780," *Estudios de Cultura Maya*, 24: 137–171.

Clark, J., M. E. Pye and D. C. Gosser (2007), "Thermolithics and Corn Dependency in Mesoamerica," in L. Lowe and M. E. Pye, eds., *Archaeology, Art and Ethnogenesis in Mesoamerican Prehistory: Papers in Honor of Gareth W. Lowe*, 23–61, Provo, UT: New World Archaeology Foundation, Brigham Young University.

Classen, C. and D. Howes (1996), "Making Sense of Culture: Anthropology as a Sensual Experience," *Etnofoor*, 9 (2): 86–96.

Comis, L. and R. Corrado (2009), "The Archaeology of Taste: Gargilius Martialis's Garum," *EuroRea*, 6: 33–38.

Farris, N. (2012), *La sociedad maya bajo el dominio colonial*, Mexico City: Artes de México, Conaculta.

Fernández Souza, L. (2016), "Grinding and Cooking: An Approach to Mayan Culinary Technology," in S. I. Ayora-Díaz, ed., *Cooking Technology: Transformations in Culinary Practice in Mexico and Latin America*, 15–28, London: Bloomsbury Academic.

Fernández Souza, L., L. Toscano and M. Zimmermann (2014), "De maíz y de cacao. Aproximación a la cocina de las élites mayas en tiempos prehispánicos," in S. I. Ayora-Díaz and G. Vargas Cetina, eds., *Estética y poder en la ciencia y la tecnología*, 107–130, Mérida: Universidad Autónoma de Yucatán.

García Barrios, A. (2017), "The Social Context of Food at Calakmul, Campeche, Mexico: Images Painted on the Pyramid of Chiik Nahb," in M. Paxton and L. Staines Cicero, eds., *Constructing Power and Place in Mesoamerica*, 171–190, Albuquerque: University of New Mexico Press.

García Barrios, A. and R. Carrasco Vargas (2008), "Una aproximación a los estilos pictóricos de la Pirámide de Las Pinturas en la Acrópolis Chiik Nahb' de Calakmul," in J. P. Laporte, B. Arroyo and H. Mejía, eds., *XXI Simposio de Investigaciones Arqueológicas en Guatemala, 2007*, 848–867, Guatemala City: Museo Nacional de Arqueología y Etnología.

Garza, M. de la, A. L. Izquierdo, M. Del C. León and T. Figueroa (2008), *Relaciones Histórico-geográficas de la Gobernación de Yucatán, two volumes*, Mexico City: Universidad Nacional Autónoma de México.

Götz, C. (2014), "La alimentación de los mayas prehispánicos vista desde la zooarqueología," *Anales de Antropología*, 48 (1): 167–199.

Hamilakis, Y. (2011), "Archaeologies of the Senses," in T. Insoll, ed., *The Oxford Handbook of the Archaeology of Ritual and Religion*, 208–224, New York: Oxford University Press.

Hamilakis, Y. (2015), "Food as Sensory Experience," in K. Bescherer Metheny and M. Beaudry, eds., *Archaeology of Food: An Encyclopaedia*, 205–206, Lanham: Rowman and Littlefield.

Hastorf, C. (2017), *The Social Archaeology of Food*, Cambridge: Cambridge University Press.

Henderson, J. S., R. A. Joyce, G. R. Hall, W. J. Hurst and P. E. McGovern (2007), "Chemical and Archaeological Evidence for the Earliest Cacao Beverages," *PNAS*, 48 (27): 18937–18940.

Kantún-Rivera, G. (2010), "Pintar escribiendo y escriber pintando. El desciframiendo de los jeroglíficos mayas," in L. Fernández Souza, ed., *En los antiguos reinos del jaguar*, 99–116, Mérida: Secretaría de Educación Pública, Giobierno del Estado de Yucatán.

Kettunen, H., and C. Helmke (2011), "Introducción a los Jeroglíficos Mayas," in *XVI Conferencia Maya Europea Copenhague 2011*. Copenhagen: Department of Indigenous Languages and Cultures, Institute for Transcultural and Regional Studies, University of Copenhagen. www.wayeb.org/resources-links/wayeb-resources/workshop-handbook/ (accessed October 5, 2018).

Landa, D. de (2001), *Relación de las Cosas de Yucatán*, Mérida: Editorial Dante.

Lentz, D. L. (1999), "Plant Resources of the Ancient Maya: The Paleoethnobotanical Evidence," in C. White, ed., *Reconstructing Ancient Maya Diet*, 3–18, Salt Lake City: The University of Utah Press.

Matos Llanes, C. M. and G. Acosta Ochoa (2016), "El patio, un estudio etnoarqueológico de actividades. Conjunción interpretativa de análisis químicos de suelos y gránulos de almidón," *Temas Antropológicos*, 38 (1): 1–30.

Mennell, S. (2005), "Taste, Culture and History," *Petits Propos Culinaires*, 78: 22–30.

Pohl, M., K. Pope and J. Jones (2000), "Base agrícola de la Civilización Maya de las Tierras Bajas," in J. P. Laporte, H. Escobedo, B. Arroyo and A. C. de Suasnávar, eds., *XIII Simposio de Investigaciones Arquelógicas en Guatemala*, 258–267, Guatemala: Museo Nacional de Arquelogía y Etnología.

Reents-Budet, D. (2006), "The Social Context of Kakaw Drinking among the Ancient Maya," in C. L. McNeil, ed., *Chocolate in Mesoamerica*, 202–223, Gainesville: University Press of Florida.

Rosenwig, R. M. (2006), "Sedentism and Food Production in Early Complex Societies of the Soconusco, Mexico," *World Archaeology, Sedentism in Non-Agricultural Societies*, 38 (2): 330–355.

Roys, R. (1972), *The Indian Background of Colonial Yucatan*, Norman: University of Okahoma Press.

Smith, M. (2006), "The Archaeology of Food Preference," *American Anthropologist*, 108 (3): 480–493.

Stoller P. and C. Olkes (1989), "The Taste of Ethnographic Things," in *The Taste of Ethnographic Things: The Senses in Anthropology*, 336–352, Philadelphia: University of Pennsylvania Press.

Stuart, D. (2006), "The Language of Chocolate: References to Cacao on Classic Maya Drinking Vassels," in C. L. McNeil, ed., *Chocolate in Mesoamerica*, 184–201, Gainesville: University Press of Florida.

Sutton, D. E. (2010), "Food and the Senses," *Annual Review of Anthropology*, 39 (1): 209–223.

Taube, K. (1989), "The Corn Tamale in Classic Maya Diet, Epigraphy, and Art," *American Antiquity*, 54 (1): 31–51.

Tokovinine, A. (2016), "'It Is His Image with Pulque': Gifts, Drinks and Political Networking in Classic Maya Texts and Images," *Ancient Mesoamerica*, 27 (1): 13–29. doi: 10.1017/S0956536116000043.

Vail, G. (2009), "Yearbearer Rituals and Prognostications in the Maya Codices and Landa's Relación de las cosas de Yucatán," in A. Gunsenheimer, T. Okoshi and J. Chuchiak, eds., *Text and Context: Yucatec Maya Literature in a Diachronic Perspective*, Bonn Americanist Studies, 47: 53–82, Aachen: Shaker Verlag.

Zimmermann, M. (2008), "Factibildad de la Paleobotánica en el Norte de la Península de Yucatán: una aproximación a los patrones des conservación de los materiales botánicos del sitio San Pedro Cholul," BA Thesis, Universdad Autónoma de Yucatán, Mérida.

Zizumbo Villarreal, D., M. García and P. Colunga García-Marín (2008), "El origen de la agricultura, la domesticación de plantas y el establecimiento de corredores biológicoculturales en Mesoamérica," *Revista de Geografía Agrícola*, 41: 85–113.

Zizumbo Villarreal, D., A. Flores Silva and P. Colunga García-Marín (2012), "The Archaic Diet in Mesoamerica: Incentive for Milpa Development and Species Domestication," *Economic Botany*, 66 (4): 328–343.

2

Gastronomy and the Origins of Republicanism in Mexico

Sarah Bak-Geller Corona
Universidad Nacional Autónoma de México

The secularization of gourmandise and the invention of *bon gout*

The emergence of the concept of gastronomy in Mexico is part of a wider process of political and social reconfiguration, marked by the period of transition from the colonial regime to the establishment of the modern state. In this context, the proposal to create Mexico's first culinary academy in 1809 established the precedent of a new kind of knowledge, a rational and systematic ordering of culinary expertise, which allowed the elites of colonial Mexico (New Spain) to conceive the country's social and political modernization. The formalization and institutionalization of culinary knowledge represents one of the first projects intended to train modern citizens in Mexico. This project happened to be based upon teaching a secularized cuisine, promoting republican principles of rationality, egalitarianism, and the common good.

Until the first half of the nineteenth century, the Castilian language did not include an alternative definition of the term "gourmandise" (*gula*) to refer to the pleasure that results from eating. However, French etiquette manuals, which had been translated and exported to the Americas since the 1820s, were already attaching a very modern meaning to the word "gourmandize." The disassociation of the notion of "gourmandise" from its strictly religious meaning in France, and its progressive connotation of good taste (*bon gout*), must be understood in terms of the medical and social transformations of the eighteenth century (Quellier 2013: 118–120). On the one hand, new scientific explanations of the digestive process called into question the Galenic dietetic theory, where foods were judged to be more or less appropriate depending on each individual's temperament (sanguine, choleric, melancholic, or phlegmatic) (Prost and Wilgaux 2006).[1] These medical advancements contributed to the emergence of a new discourse, which explicitly admitted that people ate for pleasure. From then on, culinary skills no longer sought to correct the nature of a food and make it more digestible for a specific temperament, but seasoning and cooking abilities pursued the goal of stimulating the appetite and satisfying the most demanding palates.

The sociologist Stephen Mennell explains that an improvement in food provisions (more stable, regular, and diverse), compared to earlier centuries, made possible a "process of appetite domestication" among the elites, leading to the culture of *gourmet* (Mennell 1987). Given that a better food supply assured the aristocratic and the bourgeois table, new codes of behavior and taste promoted social distinction. *Le bon gout* didn't depend on the quantity of dishes but on the quality of food and wine. The word "gourmandise" was gradually losing its sinful essence and took on a decidedly positive meaning. This began with the publication of the *Gourmands' Almanac* ([1803–1812] 1978) by Grimod de La Reynière, and accelerated following the publication of Brillat-Savarin's book *The Physiology of Taste* ([1826] 1930), considered the founding text of gastronomic literature. Brillat-Savarin explained to his French readers, who understood the word "gourmandise" only in its moral and religious sense, that this "stimulating, reasoned and regular predilection for that which flatters the palate" did not run contrary to Christian dogma: "In terms of morality, it's an implicit submission to the orders of our Creator who, having commanded us to eat in order to live, invites us to do so through the appetite, supports us through it through flavor, and recompenses us for it through pleasure" ([1826] 1930: 157).

However, Hispano-American etiquette manuals, and the Castilian language in general, refused to adopt the modern sense of "gourmandise." This word continued—and still continues today—to refer solely to one of the seven deadly sins.[2] It was not until the mid-nineteenth century that the modern, positive sense of "gourmandise" became accessible in the Castilian language, albeit through the translation of another, equally new French term instead: "*gastronomie*"[3] (*gastronomía*). In 1852 two events encouraged the emergence of words for "gastronomy" and "gourmet" in the Hispano-American context: first, the publication of the Real Academia's tenth edition of the Spanish dictionary; and second, the first Mexican edition of Brillat-Savarin's book *The Physiology of Taste*.

Although the notion of gourmandise remained closely linked to a moral and religious understanding of dietary behavior until the mid-nineteenth century, we can nevertheless observe a gradual evolution in the way that the pleasure of eating was understood and represented. The etiquette manuals that began circulating in Mexico from the 1820s onwards evidence this. In fact, the first of all these manuals, the *Tratado de las obligaciones del hombre* by Juan de Escoiquiz ([1819] 1826), was already based on a new analysis of the moral and social code that governs the act of eating. From then on, although teachings continued to promote traditional Christian values, they stopped emphasizing the impacts that "disordered passion for eating and drinking" had upon the soul (Escoiquiz [1819] 1826: 30). What mostly concerns the author of this manual are the harmful effects of gourmandise on an individual's material situation and social rank. This change, which appears to superimpose liberal logic onto Christian morality, is a good illustration of the new preoccupations of modern Hispano-Americans, which no longer corresponded with those of the Church—or at least not solely.[4] The effects of irrational food consumption and associated expenditure were still seen as negative, but no longer framed in terms of morality. Instead, the concern now related to the finances and health of the consumer, because excess "damages health and burdens the body and the soul, diminishing the ability to move; a too strong passion for eating and drinking leads many people to squander their wealth and reduce themselves to penury" (Escoiquiz [1819] 1826: 30). Although a marked weakness for good food remained the subject of criticism, the codification of dietary behaviors and the professionalization of culinary knowledge were seen as positive tools of Republican civil society, and increasingly became a political and social priority within official objectives. This can be clearly seen in the proposal to establish Mexico's first culinary academy, published in the *Diario de México* in December 1809 (Tepalcate).

The codification and democratization of culinary knowledge

On December 14, 1809 the *Diario de México*, "Tepalcate"[5] (the parish priest of Santa Cruz, writing under a pseudonym)[6] published an article entitled "*Culinaria*." This article put forward a proposal for the "public good," which was intended to address New Spain's backwardness compared to more "civilized" societies. The author was concerned by the irrationality and confusion surrounding culinary tasks in the country, and proposed the creation of several culinary academies. These would teach students to develop "culinary skills," which would be translated into civic virtues: order, equality, productivity, and patriotism.

With this objective of formalizing and institutionalizing culinary knowledge, Tepalcate constructed cuisine as "classic" and universal. The origins of the word *culinaria* (from the Latin *culinarius*) offer links to the language of the ancient Romans: "I take this word from the Latin, and so no grammarian would hesitate to give it a passport." He then produced a culinary mythology and pantheon, featuring classical figures portrayed as pseudofounding fathers of the art of cuisine. Figures invoked include Eglon,[7] Vitellius,[8] Lucullus,[9] and Elagabalus,[10] and these figures' negative images as the tyrants of historical accounts are ultimately erased by their reputation as great gourmands. Tepalcate thus definitively established cuisine as a form of knowledge that had been recognized and appreciated since ancient times, giving his project a historic and cosmopolitan dimension. However, he never lost sight of his principal goal, which was to adapt the classical model to the essential needs of the country's inhabitants. This is why Tepalcate suggested that the Old Testament quote "Erat autem Eglon crasus nimis" (Eglon was extremely fat) should become the motto of the culinary academy, and he specifies that this quote should be engraved in Castilian, "so that it can be understood by cooks and kitchen boys" (Judg. 3:17). The parish priest of Santa Cruz evidently aimed to involve society's working classes in the project of professionalizing culinary tasks, which would require circumspect and moderate behavior toward food.

With a view to integrating culinary knowledge into New Spain's modernizing project (Bak-Geller 2015, 2018), Tepalcate recommends that a dictionary of cuisine should be produced, in order to make the most famous recipes accessible to the greatest number of aspiring chefs, irrespective of their social or ethnic origins. This would be an exhaustive and encyclopedic compendium giving access to an unequalled collection of recipes. The recipes would be ordered by method of preparation ("stews, pot-au-feu, roasts, pastries, jellies, sauces, etc."), and would also recount the history of the dishes "from the origins of nutrition to the present day." Tepalcate's idea of a dictionary did not become reality until 1845, when the editor Mariano Galván published the *Diccionario de cocina o Nuevo cocinero mexicano*. The advantages of this format, which Galván lays out in the prologue, are surely the same as Tepalcate had in mind half a century before. The dictionary promised to solve "the difficulty experienced by some, who are little versed in the handling of books," and who were confused by "the multitude and variety of its articles, [with] frequent loss of time and patience" (Galván [1831] 1845: prologue).

Tepalcate also proposed establishing a method of selection based on exams and prizes, particularly prizes for "outstanding genius [and] chefs-d'œuvre," constituting the highest recognition of culinary skills and talent. This system of prizes and recognition for individual merit, as opposed to systems of rank and position that characterized the old order, would encourage a process of social mingling and leveling in the Kingdom of New Spain. In this new democratizing logic, personal skills were valued above birth and rank.

Thanks to these measures, the job of chef would become a legitimate profession, validated by an institution and a body of specialists. The Culinary Academy had the task to form the first professional body, followed by future generations of chefs. However, women were granted little or no role within this new conception of culinary practice. Tepalcate relegated female chefs to a "species of the third-order" and lowest rank; he invoked the intellectual limitations of the feminine mind, which apparently manifested in irrational and capricious conduct when women cooked. Women were considered incapable of adapting to concepts of norms, order, and discipline, and were therefore seen as unsuitable to become chefs useful to society. The same reasons for excluding women from the project of culinary academies were used to justify their disqualification as possible citizens.

At the end of the first decade of the nineteenth century, in the midst of the political crisis that threatened the monarchy, the culinary academies expressed a project of modern citizenship. Citizenship emerged as a political community made up of equal, rational, productive, and useful individuals. In this ideal of political community, professionals in the field of cuisine would be one of the groups responsible for teaching republican behavior to the country's inhabitants. The professionalization of culinary tasks thus allowed the regulation of culinary activity in public spaces: on the one hand, through the assessment of the culinary skills of any individual who wished to devote themselves to selling their cooking; and, on the other hand, through the establishment of regulations aimed at all food businesses in a town:

> The Academy must create regulations for inns, taverns, cafes, bakeries and other stalls [...], so that [food] will be served by authorized individuals who hold a suitable official qualification, and so that these establishments will respect the standards and rules prescribed by the regulations. (Tepalcate 1809)

Finally, Tepalcate proposed the creation of a code of cuisine that would train students in the main rules and criteria of the culinary art (1809: 677–679). The first Code of Cuisine in Mexico claimed to establish the foundations of a uniform and rational system of knowledge, endowing it with "principles, rules and laws for its proper usage and splendor, such as it deserves." The author thus sought to put an end to the arbitrary nature of tastes and methods that characterized the kingdom's diverse cuisines. For several years previously, the theme of standardization and homogenization of culinary practices had taken a central place in the arguments published by newspapers and gazettes, in which the matter was a subject of debate. One of these voices was Dr. Desiderio de Osasunasco, author of an article entitled "Observations on the preparation and uses of chocolate" published in the *Gazeta de México* in 1789. He expressed his discontent over the lack of rigor in the methods used and proposed some parameters to make them more uniform. Yet the problem of food norms arose in different terms for Tepalcate, for whom perfecting dishes was not a major concern. In contrast, he felt that the essential, necessary duty was to create a regulated corpus of culinary knowledge that would be the sole reference point for cooks in New Spain. A culinary practice subject to general laws and principles, which applied to all without any distinction of race or social class, would act as a pillar of the new society of modern citizens.

Science and cuisine, a formula for the republic

Tepalcate may have drawn inspiration from the gastronomic writings of Grimod de La Reynière, who was the very first to discuss cuisine within a legitimate process of reflection, critique, and debate. La Reynière was a writer, journalist, and theatre critic, and in 1808 he published his famous *Manuel des Amphitryons*, a foundational text of gastronomic literature. In this work, La Reynière also expresses the desire to institutionalize culinary knowledge, proposing that *lycées* should have chairs of gastronomy (La Reynière [1808] 1983: xxxiii). On this point, the concerns of La Reynière and Tepalcate converge; however, given that these two writings were published in close succession (hardly a year apart), it seems unlikely that Tepalcate could have had access to La Reynière's book before drawing up his own proposal. However, six years earlier, in 1803, La Reynière had published the first edition of his *Gourmands' Almanac*, which had been an immediate success. It is highly probable that Tepalcate, when he published his article in the *Diario de México* in 1809, had already become aware of the work in question. At first glance, the similarities between the two documents raise questions about the contexts that might give rise to this new enthusiasm for cuisine, on both sides of the Atlantic. At the same time, this comparison allows us to observe differences between these two texts, which highlight the originality of the proposals conceived in the parish of Santa Cruz.

The *Almanac* had just one objective, which appears several times throughout the work: to serve as a guide for the empire's *parvenus* who were unaware of old-fashioned codes of manners. For Grimod de La Reynière, a direct descendant of an aristocratic family, the fault for this lay with the revolution, which had destroyed old culinary traditions and did not know how to replace them:

> The revolution changed all that: fortunes passed into other hands, new customs replaced old [...]. What became of these delicious suppers, where all the most pleasant things which the Court, the Town and the Arts had to offer were gathered together in sanctuaries of opulence; which were governed by this equality without which all pleasures are naught, where rank, birth, intellect and fortune did not appear save to blend together with sweetness, taste and graces; where no superiority made itself felt, because worldly customs had taught restraint in one's pride; and where the beauty of the day, the fashionable poet, the all-powerful minister and the favored courtesan appeared indistinguishable under the flag of true liberty. Alas! They disappeared in this revolutionary torrent that overwhelmed and ravished all. (La Reynière [1803–1812] 1978: 84–85)

La Reynière thus assigned himself the task of restoring the cuisine and the good manners of days gone by. His concern was safeguarding and consolidating the customs of the *Ancien Régime*, which he idealized. On the other hand, without wanting to slip into too reductionist a vision, Tepalcate's intentions might be defined as diametrically opposed to La Reynière's: what La Reynière viewed as an act of nostalgia, a "return to everything that was eminently typical of the customs of the French nation" (La Reynière [1808] 1983: 92), Tepalcate saw as a project of social renovation, an opportunity to lay the foundations of a new social order, governed by the universal principles of science and reason. Tepalcate continually looked to a more promising future, believing that traditional customs were out-of-date and that space must be made for a new society better adapted to a more advanced era.

Tepalcate defended the idea that although cooking is an everyday activity, it requires a high level of qualification and knowledge, just like art and science.[11] The question of the status of cooking in the field of human knowledge was not new. Since ancient Greek times, the link between cooking and art had been a subject of debate. For Plato, who was one of the first to consider this relationship, gastronomy could not be classed as an art; it was considered an "ignoble" form of knowledge, which should be in the service of medicine and gymnastics, the true arts of the body: "and I say that [gastronomy] is not an art, but a knack, because it has no account of the nature of whatever things it applies by which it applies them, so that it's unable to state the cause of each thing. And I refuse to call anything that lacks such an account a craft" (Plato 1997: 808–809). "Gastronomic adulation" resembled rhetoric, argued the philosopher, in the sense that they both "masked" the truth of things in the sole aim of giving pleasure (Plato 1997: 809; 861–863). That said, Plato himself recognized that most people paid more attention to cooks than to doctors when it came to dietary advice (Plato 1997: 808).

In an effort to systematize culinary knowledge and give it a higher status, Tepalcate proposed the establishment of a classification of different culinary techniques; these would be divided into several branches, such as patisserie (which would then be considered a discipline in its own right), or the new chemical science of foods.[12] This science could lead cooks to research new "raw materials, compositions, and transformations." The distinctive feature of this new culinary science would be to democratize access to food, through the discovery of "the secret for making all or certain products of the three kingdoms of nature—which had, until these obscure times, never been made consumable—not only edible, but digestible and pleasant." Afterward, the culinary art, enriched with the most rigorous of scientific methods, took shape as a true discipline in the service of the social and material development of Mexican society.

The modern dining table: Progress and nostalgia in gastronomic discourse

A good proportion of New Spain's intellectual elite shared this full confidence in the progress of culinary arts and good dining etiquette as a means of laying the foundations for modern citizenship. In 1809, the same year that Tepalcate's proposals appeared, the *Diario de México* published another enthusiastic article, in which the author interpreted the recent observable changes in the country's cooking as signs of clear social progress, stating that "cooking today is more delicate, refined and wholesome than it was forty years ago" (*Diario de México* 1809).

The 1850s was a time of further celebrations of advancements in cooking, a field that promised to occupy a greater and better place in the new republican imagination. A particular genre of books testified to this confidence in the future of the new republic: cookbooks. In the prologues and the etiquette codes that were often included at the end of these manuals, the authors introduced new modes of dining behavior that promised to transform Mexico into a modern and democratic society. Galván, the editor of the *Nuevo cocinero mexicano* ([1831] 1845), the first Mexican cookbook, speaks of the role of the dining table as an ideal place to "reinforce social ties," particularly for nonfamilial relationships. The modern dining table was seen as a place where equality was the prevailing principle, where individuals' "rank, birth, talent and fortune" became indistinguishable "under the banners of true liberty" (Galván [1831] 1845: prologue and appendix).[13] A clear contrast was being established vis-à-vis the *Ancien Régime*'s rules of sociability. The colonial elites' experience at the table, represented in terms of status, class, and "character," is visible in the first Hispano-American novel, *El periquillo sarniento* (1816) by José Joaquín Fernández de Lizardi, published four years before Mexican independence. In one of the episodes, the main character embodies the colonial social stratification by assigning himself a seat according to his social status: "It was time to eat, they set the table, and we all sat according to the class and character of each one" (Fernández de Lizardi [1816] 1842: 62).

Far from being considered trivial, cooking and the sociability of dining have often been considered a true measure of civic life. In classical political discourses, notably those of Cicero, we find the first image of the dining table as a place for teaching republican values (equality, liberty, virtue, and the common good). For Cicero, the convivial dimension of dining was paramount, as he noted that the Latin word *convivium* was more appropriate than the Greek vocabulary, which evoked only the material aspect of the meal: food and beverages (Assouly 2016: 191). The seventeenth-century German philosopher Johannes Althusius, whose work sought to promote civic behavior, devoted a chapter to the virtue of temperance and moderation through the art of civil eating (Van Gelderen and Skinner 2002: 206). The gastronomic language in the early nineteenth century is filled with classical references to the dining table as a privileged site of civility.

José María Luis (1794–1850), an active proponent of liberal thought in the early days of the Mexican republic, emphasized the importance of *trato social* (social treatment) in the advancement of national culture, particularly where the dining table was concerned. He illustrated this belief through the description of a "Mexican table" possessing all the attributes of the great Western civilizations:

> But the most remarkable thing about the advancement of Mexican civilization can be seen in sociability, or that which comprises the attractions of social treatment: the fairer sex, the clothing, the attendees, the promenades, the distractions, and the *pleasures of the Mexican table* have undergone radical changes or made considerable progress. (Mora [1836] 1997: 48; emphasis mine)

Despite the prevailing confidence in the development of the art of cuisine as a civilizational tool, there were of course some who saw it as a reflection of society's deterioration. This pessimistic position highlighted the disadvantages of the boom in cuisine and the new social model of dining, for three principal areas: health, economy, and morality. The editor of one of Mexico's periodicals, José Antonio Alzate (1737–1799), is a worthy spokesperson for this classical discourse on the abandon with which spices were used in this ideal republican diet. Following the Platonic tradition, which postulates the advantages of an austere and frugal diet for citizens,[14] Alzate maintains that the ascetic cuisine of their ancestors was much more wholesome than that of his contemporaries, which he condemns for its excessive use of spices:

> I think that spices, due to their acrimony, should be kept by apothecaries, where their medical uses can be extracted by the skilled hand of a professor. Some will say that this is trying to return centuries into the past: this is what we should do with food: we should re-establish the noble customs, the simplicity in food provision, which kept our ancestors healthy and robust. (Alzate [1790–1792] 1831: 238–239)

The innovations arriving at dining tables in New Spain even perturbed some of the most reformist men of the age, such as the journalist José Joaquín Fernández de Lizardi (1776–1827), who observed with some displeasure the great novelty of the moment: the introduction of the menu, showing the dishes available and their prices, which had been fixed in advance. Fernández de Lizardi was used to going into inns and taverns and ordering food without any restriction on his choices, and to having dishes "prepared to each individual's taste" (Osasunasco 1794). In the menus introduced by restaurants in the capital, he thus saw only an opportunity for swindling. This journalist did not appreciate the apparent advantages of the new system, in terms of the quality of the food prepared and the accuracy of the prices, who instead explained how the patrons of restaurants would profit from this method by arbitrarily imposing unreasonably high prices upon their customers (Fernández de Lizardi [1812–1814] 1987: 16). Dining customs were changing, and Fernández de Lizardi demonstrated, through his reproach to restaurateurs, that he was part of a generation which ill understood not only the new eating practices but also the social and political ordering that lay behind the menu. The standardization of food choices revealed an idea of political community composed of equal individuals, undifferentiated by class, rank, or cultural preferences.

The introduction of new technologies in cuisine in the early nineteenth century, particularly the invention of the corn mill, provoked further expressions of pessimism and confusion among the educated classes of Mexican society. A study intended to promote the use of the corn mill in the country explained the benefits in time and energy that came from replacing the *metate* with the new machine for milling corn; and yet, at the same time, the author highlighted a decline in the flavor of the dough obtained by this method (Azcárate [1839] 1902: 86). He regretted that the taste of the modern *tortilla* was no longer the same as that which had been prepared with the grindstone, observing that the "dough made by the mills is not as flavorful as that produced *by hand and by arm*" (Azcárate [1839] 1902: 86).

The social and political elites of the early nineteenth century widely shared the idea that culinary science and industry brought improvements to Mexican society. However, in response to the enthusiasm provoked by the idea of breaking with old customs, came the first displays of nostalgia for culinary knowledge that was now being relegated to the past. In this back-and-forth between proposals for constructing a new society, on the one hand, and the regret for an idealized past, on the other, we can thus see the spirit of modernity driving the gastronomic language of actors engaged in constructing the new Mexican republic.

Notes

1 Hippocratic-Galenic theory was based on the existence of four humors that governed the human body (blood, phlegm, black bile, yellow bile) whose balance indicated the state of health and temperament of each person. Their balance could be corrected at all times with specific diets.

2 Brillat-Savarin himself argued that the modern word "gourmandise" could not be translated. He was proud of this fact, and suggested that the French word should be retained in other languages ([1826] 1930: 167).

3 The term "gastronomy" appeared for the first time in the 1801 poem "La Gastronomie ou L'homme des champs à table" (Gastronomy, or The man of the fields at the table) by the lawyer Joseph Berchoux. This new term was a French translation of the Greek word used by the poet Archestratus (fourth century BCE), who formed it from two words: stomach (*gastros*) and laws that govern (*nomos*). The poem by Archestratus of Gela, conceived as a tour of the Mediterranean's culinary specialities, was very popular in its day, according to Athenaeus's comments in *Deipnosophistae* ("Sophists at Dinner," second to third century CE).

4 A similar semantic development can be found in most Spanish school textbooks of the period, see Guereña 1997: 51–58.

5 The word *tepalcate* is of Nahuatl origin and refers to the fragment of a terracotta receptacle, generally destined for culinary use. Given the pejorative meaning of

this word in popular language, we interpret this choice of pseudonym as a sign of humility on the part of the priest.

6 It is impossible to know the exact location of this parish, given that the name Santa Cruz was one of the most commonplace names in New Spain.

7 According to the Old Testament, Eglon, the king of Moab, was so fat that the dagger that Ehud used to kill him, and thus free the Israelites, disappeared into the fat of his stomach (Judg. 3.21–25).

8 In *The Lives of the Twelve Caesars*, Suetonius recounts how this Roman emperor (14–69 CE) became one of the greatest gourmands of antiquity.

9 Lucullus was the conqueror and proconsul of the Roman province of Asia (115–57 BCE). His biography was written by Plutarch in the *Lives of the Noble Greeks and Romans*, which includes descriptions of the sumptuous banquets organized by Lucullus. English language has kept the word "Lucullan" to refer to that which is produced with abundance and opulence, particularly in relation to cuisine; and the French have a dish named *escalope Lucullus*.

10 Also called Heliogabalus. Elagabalus was a Roman emperor (218–222 CE). Legend has it that Elagabalus killed his guests during a dinner by suffocating them under rose petals that he had falling from a false ceiling. This story made an impression in Hispanic culture, where the word *heliogábalo* is used to describe a person dominated by gourmandise.

11 Until the mid-eighteenth century, in France's general system for cataloguing books, cookbooks were classed under "Medicine." From 1750 onwards, the expression "the art of cooking" appeared in the bibliographic literature, and cookbooks began to be placed in the "Arts" rubric, thus emerging from medicine's purview. Cookbooks thus came to share the same status as books on horse riding, fencing, dancing, and hunting, all firmly aristocratic leisure pursuits (Viardot 1996: 139–140).

12 A detailed study of chemical theories of food in eighteenth-century France can be found in Spary 2014.

13 Similar discourses can be found in the other new American republics. See, for example, the national cookbooks of Brazil (*Cozinhero Nacional,* 1877) and Cuba (*Nuevo manual del cocinero cubano y español*, 1857). The anniversary of Colombian independence, it was said, was celebrated in 1838 in Bogotá with a banquet where men of different social origins (artisans, farmers, soldiers, and literary men) shared tables "under the doctrine of equality" (Restrepo Manrique 2012: 223). At White House dinners, Thomas Jefferson instituted the "pell-mell" table seating system, which required all to be treated equally as citizens of equal rank. The British ambassador rejected Jefferson's conception of the democratic and refused to attend dinners at the White House (DuPuis 2015: 25).

14 The "wholesome" city, as opposed to that which is "infected" and "corrupt," is characterized by the simple diet of its citizens: tarts and breads made from barley and wheat flour, salt, olives, cheese, onions and boiled vegetables, figs,

peas, fava beans, and toasted acorns (Plato 1997: 1011–1012). In the eighteenth century, frugality is recognized as one of the main doctrines of republicanism. The *Nuevo vocabulario filosófico-democrático* (Thjulen [1799] 2006), a satirical and counterrevolutionary book, notes that the concept "frugality" is one of the fundamental republican dogmas. Nevertheless, Thjulen, being a furious anti-revolutionary, mocks the republicans and portrays them as gluttons and epicureans ([1799] 2006: 70, 71).

References

Alzate, J. A. ([1790–1792] 1831), *Gacetas de Literatura de México*, Puebla: Oficina del Hospital de San Pedro, a cargo del ciudadano Manuel Buen Abad.

Assouly, O. (2016), *Les Nourritures de Jean-Jacques Rousseau. Cuisine, goût et appétit*, Paris: Classiques Garnier.

Azcárate, J. M. ([1839] 1902), *Noticias estadísticas sobre efectos de consumo en México, correspondientes al quinquenio de 1834 a 1838*, Mexico City: Imprenta de la Águila, reprinted in *El Agricultor Mexicano*.

Bak-Geller Corona, S. (2015), "Wheat *versus* maize: Civilizing Dietary Strategies and Early Mexican Republicanism," *Journal of Interdisciplinary History of Ideas*, 4 (8): 3–25.

Bak-Geller Corona, S. (2018), "Comida, civilización y república. El pensamiento de Rousseau en la conformación de una dieta patriótica en México," in G. Entin, ed., *Rousseau en Iberoamérica: del reformismo borbónico a las revoluciones de independencia*, Buenos Aires: SB Paradigma inicial.

Brillat-Savarin, J. A. ([1826] 1930), *La physiologie du goût* [The Physiology of Taste], Paris: Éditions du raisin.

Diario de México (1809), *Diario de México*, vol. 11: 677–679, April 4, 1809.

DuPuis, E. M. (2015), *Dangerous Digestion: The Politics of American Dietary Advice*, Oakland: University of California Press.

Escoiquiz, Juan de ([1819] 1826), *Tratado de las obligaciones del hombre en la sociedad*, Burdeos: Imprenta de Dn Pedro Beaume.

Fernández de Lizardi, J. J. ([1816] 1842), *El periquillo sarniento*, vol. 1, 4th edn., Mexico City: Librería de Galván.

Fernández de Lizardi, J. J. ([1812–1814] 1987), *El Pensador Mexicano*, vol. 4, Mexico City: Centro de Estudios de Historia de México Condumex, fac-similé.

Galván, M., ed. ([1831] 1845), *Diccionario de cocina o Nuevo Cocinero Mexicano en forma de diccionario*, Mexico City: Imprenta de I. Cumplido.

Guereña, J.-L. (1997), "La transmission des codes sociaux dans l'espace scolaire en Espagne au XIXe siècle," *Romantisme*, 96: 51–58.

La Reynière, G. de ([1803–1812] 1978), *Almanach des Gourmands*, in *Écrits gastronomiques*, Paris: 10/18.

La Reynière, G. de ([1808] 1983), *Manuel des Amphitryons, contenant un traité de la dissection des viandes à table, la nomenclature des menus les plus nouveaux et des éléments de politesse*, with a preface by Misette Godard, Paris: A. M. Métailié.

Mennell, S. (1987), *Français et Anglais á table du Moyen Âge à nos jours*, Paris: Flammarion.

Mora, J. M. L. ([1836] 1997), *El carácter de los Mexicanos*, Mexico City: Fondo de Cultura Económica.

Osasunasco, D. de (1794), "Observations on the Preparation and Uses of Chocolate," *Gazeta de México*, July 21, 1794.

Plato (1997), "Gorgias," trans. D. J. Zeyl, in J. M. Cooper and D. S. Hutchinson, eds., *Plato: Complete Works*, 791–869, Indianapolis, IN: Hackett Publishing Company.

Prost, F. and J. Wilgaux, eds. (2006), *Penser et représenter le corps dans l'Antiquité*, Rennes: Presses Universitaires de Rennes.

Quellier, F. (2013), *Gourmandise. Histoire d'un péché capital*, Paris: Armand Colin.

Restrepo Manrique, C. (2012), *La alimentación en la vida cotidiana del Colegio Mayor de Nuestra Señora del Rosario, 1653–1773/1776–1900*, Bogotá: Ministerio de Cultura.

Spary, E. C. (2014), *Eating the Enlightenment: Food and the Sciences in Paris, 1670–1760*, Chicago: University of Chicago Press.

Tepalcate [pseud.] (1809), "*Culinaria*," *Diario de México*, December 14, 1809.

Thjulen, L. I. ([1799] 2006), *Nuevo vocabulario filosófico-democrático indispensable para todos los que deseen entender la nueva lengua revolucionaria*, Valladolid: Maxtor.

Van Gelderen, M. and Q. Skinner (2002), *Republicanism and Constitutionalism in Early Modern Europe*, vol. 1, Cambridge: Cambridge University Press.

Viardot, J. (1996), "Livre de cuisine et collectionneurs français," *Papilles*, 10/11: 137–144.

3

Alcohol Consumption Patterns among Different Social Groups during Yucatán's Gilded Age

Héctor Hernández Álvarez and Guadalupe Cámara Gutiérrez
Universidad Autónoma de Yucatán

Introduction

Yucatán's Gilded Age (1873–1927) was characterized as a period of technological innovation and economic wealth that locates the peninsula's entrance into a globalized market system as a result of henequen exploitation (Wells 1985; Wells and Joseph 2011). However, the same period also exhibited remarkable social inequality, especially when contrasting the living conditions of members of the capitalist elites with that of the impoverished indigenous workers. This situation brought about a series of social, economic, and political problems—in particular, the alienation of peninsular farmers from their land imposed by henequen monoculture that forced them to serve as legally sanctioned indentured labor within the haciendas (Ortiz Yam 2013).

The historical record shows that the unrestrained consumption of alcohol, among the rich and the poor, was one of the main consequences of the industrialization and economic bonanza on the Yucatán Peninsula at the end of the nineteenth century (Cámara Gutiérrez 2009; Fallaw 2002). However, the consumption patterns of alcoholic beverages were very different between social classes. While elites imported an array of wines, beers, and liquors, among the indigenous population the consumption of locally distilled aguardiente (rum) became generalized. In fact, there is evidence that suggests that aguardiente was used as a coercive tool that kept the workforce indebted to the haciendas (Cámara Gutiérrez 1995; Peniche Rivero 2010).

Taking into account historical documents and different types of material cultural evidence, in this chapter we analyze the distinct consumption patterns of elites and peons who lived and worked at Hacienda San Pedro Cholul during the dawn of the twentieth century. Through the study of documentary sources covering the importation and consumption of alcoholic beverages in the region, and the analysis of a collection of glass shards and bottles that kept alcoholic beverages, which were recovered from the aforementioned hacienda, we show how different patterns of alcohol consumption served to draw lines of inclusion and exclusion between the period's different social groups.

The anthropology of alcohol consumption

Over the last decades, alcohol consumption has been studied by anthropologists from different perspectives (Ayora-Diaz and Vargas-Cetina 2005; Dietler 2006; Heath 1987a, 1987b). There are different studies which distinguish alcohol consumption patterns related to different social groups, identify how different ways of serving alcoholic beverages are used to create social boundaries, and seek to understand the different functions and behaviors related to drunkenness (Barrows and Room 1991; Douglas 1987). Generally, anthropologists have shown interest in unveiling the complex meanings different societies attribute to alcohol consumption (Dietler 2006; Heath 1987a; Wilson 2005).

Historical archaeology, on the other hand, examines the social forces and symbolic aspects that have shaped alcohol consumption patterns in the modern world. Archaeologists have been preoccupied in establishing the types of alcoholic beverages that groups consumed, illustrating the manner in which social distinctions based on class, race, or gender promoted differing consumption patterns (Smith 2008). Historical archaeologists in particular have conducted research on topics such as the relation between alcohol and foodways, or drinks and sociability (Karskens 2003). In addition, scholars active in the field are looking for the places of consumption—taverns, saloons, and bars (Eckholm and Deetz 1971; Noël Hume 1968). Other aspects that have received considerable attention include the material culture associated with alcohol consumption, resulting social relations, the survival of Old World traditions, the relation between alcohol and health, as well as its role as an anxiety inhibitor (Bittmann and Alcaide 1984; Fontana 1968; Kent 1983; Smith 2008).

Regarding historical research, historians have examined three foci of alcohol consumption: first, historians have documented the sociocultural context as well as biological, psychological, and social effects of alcohol usage (Bruman 2000; Eber 2000; Heath 1976, 1987a); second they have paid great attention to the production and consumption of intoxicating beverages and their relation to the state's control over aspects such as sales, prohibition, monopolies, taxation, crime, and rebellion (Cámara Gutiérrez 2007, 2009, 2010; Carey 2012; Fallaw 2002; Miranda Ojeda 2010; Torras Conangla 2007); and a third focus is the examination of the construction of alcohol demand as a cultural practice with profound political and economic implications (Dietler 2006; Jennings and Bowser 2008; Schwartzkopf 2012). This latter focus acknowledges the multiple links that exist between consumption patterns and the reproduction of social relations within families and communities.

The production, distribution, and consumption of alcohol during the Gilded Age

On the Yucatán Peninsula, the commercial production of aguardiente during the first half of the twentieth century happened synchronously with the growth and expansion in sugar production and cane alcohol distillation. The commerce of liquor was distributed throughout southern and central-eastern Yucatán, as well as northeastern Campeche. The main distribution centers were located at the towns and cities of Tekax, Ticul, and Peto in the south; Mérida, Motul, Espita, and Valladolid in the central-eastern zone; and Campeche, Seybaplaya, Hecelchakán, Calkiní, Tenabo, and Villa del Carmen in northeastern Campeche. These regions constituted the main options for access to cane derivatives for the peninsula's leading urban centers: Mérida and Campeche. At that time, Mérida was the focal point of commercial activity on the peninsula, and Campeche was the main port both for exports and imports (Cámara Gutiérrez 2009: 247). On a regional scale, the distillation of aguardiente grew and extended during the first half of the twentieth century, chiefly between 1821 and 1852 (Cámara Gutiérrez 2009).

A second phase of development occurred between 1853 and 1875, when the industry consolidated and commerce increased and diversified. In this context, since the first half of the nineteenth century, a small group of retail traders consolidated around the sale of liquor at small communities and haciendas. During the second half of the nineteenth century, even when no new areas of influence were opened to liquor traders, a large number of people became everyday consumers (Cámara Gutiérrez 2009: 247–248). By 1875, Yucatán's social scenario had become troublesome. The rise in aguardiente commerce and consumption expanded over the entire peninsula. The first social symptoms of social discontent appeared, and so did the violence attributed to excess consumption. The state's attempts to regulate commerce and consumption met with very limited results (Cámara Gutiérrez 2009).

Starting with the second half of the nineteenth century, the trade routes for aguardiente distilled from local sugar cane helped define a clear-cut geography of consumption. One large area stood out: the henequen plantations zone, where indigenous populations represented the main consumers. Here, the sales, distribution, and consumption of local aguardiente turned into an element of coercion and peonage by indebtedness (Hernández Álvarez and Cámara Gutiérrez 2017). Compared to the locally produced cane liquor, which was almost entirely consumed by the indigenous population, the elites and some of the inhabitants of larger cities consumed nationally as well as internationally imported wines and liquors.

The statistical analysis of national and foreign imports registered since 1895 at the port of Progreso,[1] displays a certain specialization in the types of goods and products acquired by companies and businesses inhabiting peninsular cities: grain, anise, sugar, casks of liquor, beer, *panela* (brown sugar loafs), and *piloncillo* (brown sugar) figure prominently. Starting in July 1896, national imports diversified significantly, with a large variety of liquors, red wine in

Table 3.1 Aguardientes, beer, liquor, and wine imported to Yucatán, 1884–1894

Years	Aguardiente (L)	Beer (L)	Liquor (L)	Wine (L)
1884	7,821	40,832	254	211,814
1885	23,582	79,726	7,750	120,993
1886	1,914	79,311	57	153,985
1887	10,730	46,713	1,256	73,784
1888	78,937	167,097	1,392	127,013
1889	115,372	504,548	8,099	936,510
1890	122,379	391,065	5,885	610,864
1891	36,703	256,930	1,872	471,569
1892	53,102	347,242	1,854	690,300
1893	81,392	324,758	2,347	867,163
1894	36,830	148,930	83	622,559
Total	**568,762**	**2,387,152**	**30,849**	**4,886,554**

Source: AGEY, Registro Público de la Propieded, Libro 1968, Boletin de Estadística, Mérida, 1896, vol. 2, year 3, p. 142.

wooden casks, bottled aguardiente, cognac, mescal, and bottled white wine. Foremost among these alcohols was the access to Saint Remy brandy.[2] Foreign imports consisted of bottled and barreled aguardiente, bottled beer, sparkling wine, casks and bottles of red and white wines, as well as a variety of liquors.[3] Especially interesting are also some goods derivative of the market of liquors and aguardientes such as disassembled wooden casks,[4] unplugged glass bottles,[5] plug cork,[6] and straw wrapping for bottles.[7]

Table 3.1 summarizes the volume or quantity of imported aguardiente, beer, liquor, and wine over the years 1884 to 1894.

The amount of beer, liquors, wines, and aguardiente consumed in Yucatán from 1880 to 1895 is presented in Table 3.2.

Table 3.2 Alcoholic beverages consumption in Yucatán, 1880–1895

	Import				Local production
Years	**Beer (L)**	**Liquor (L)**	**Wine (L)**	**Aguardiente (L)**	**Aguardiente (L)**
1880					1,012,027
1881					1,054,309
1882					1,509,624
1883					1,293,907
1884	40,832	254	211,814	7,821	917,463
1885	79,726	7,750	120,993	23,582	970,528
1886	79,311	57	153,985	1,914	967,613
1887	46,713	1,256	53,784	10,730	1,349,109
1888	167,097	1,392	127,013	178,937	1,544,859
1889	504,548	8,099	936,510	115,372	1,553,472
1890	391,065	5,885	610,864	122,379	1,497,161
1891	256,930	1,872	471,569	36,703	1,230,876
1892	347,242	1,854	690,300	53,102	1,449,948
1893	324,758	2,347	867,163	81,392	1,558,807
1894	148,930	83	622,559	36,830	1,606,616
1895					1,507,096
Total	**2,387,152**	**30,849**	**4,886,554**	**568,762**	**21,022,415**

Note: Data on imports was taken from official documentation, missing the years 1880, 1881, 1882, 1883, and 1895. Regarding local production data, it is taken from the annual output of aguardiente distillation. There is no exact data available dealing with the quantities resulting from aguardiente production for Yucatán.
Source: AGEY, Registro Público de la Propieded, Libro 1968, Boletin de Estadística, Mérida, 1896, vol. 2, year 3, p. 142.

As mentioned above, Mérida and Campeche were the two main commercial destinies for distilled, fermented, and natural drinks, either produced locally or imported nationally or internationally. However, throughout the second half of the nineteenth and the first half of the twentieth century, other cities and important towns included Tekax, Ticul, and Peto to the south, Valladolid, Tizimin, and Espita to the east, and Motul in the central region. These municipalities provided nearby populations and, specifically, hacienda workers, with locally produced cane liquor. For example, the amount of aguardiente produced in Espita surpassed local demand and was sold at markets in Temax, Cansahcab, and Motul, as well as to coastal populations and in the state capital, Mérida.[8] In Mérida, trading companies such as "El Elefante," located at the crossing of 65th and 48th Street, offered a huge variety of aguardientes, wines, and fine liquors. In 1859, the inventory and stock of the business owned by D. Roque Milán and D. Esteban Martínez included rubrics such as those listed in Table 3.3.

Table 3.3 Inventory and balance of a commercial house in Mérida, Yucatán (December 17, 1859)

Article/product	Total appraisal (Pesos)
16 pounds bitter	77.00
12 pounds bitter	341.25
Aniseed of 8	247.50
Anisadillo of 4	64.00
Anisette	56.25
Aniseed	32.40
Honey anise	15.00
English black bottles	48.50
French bottles	52.50
Champagne bottles	7.00
Irish white bottles	14.00
Moscatel wine bottles	5.00
Beer bottles	7.12 ½
Eagle green bottles	15.00
Oak barrels	2.00
Empty botellones	0.12 ½

Table 3.3 (*Continued*)

Article/product	Total appraisal (Pesos)
Empty barrels without lid	6.50
Can bottles	0.25
Glass jars	6.50
New demijohns	33.75
Demijohns in good shape	168.75
Used demijohns	92.00
Glass covered demijohns	32.00
Gin on demijohns	32.00
Islander demijohns	140.00
Regular habanero	81.60
Habanero	105.00
Veracruzano habanero	63.00
22 degrees habanero	54.00
Country habanero	56.00
Half white bottles	8.00
Half black bottles	2.00
White mistelle	3.00
Half barrels in good shape	57.50
Half white barrels	12.00
Half black jars	4.18 ¼
Half can jars	0.56 ¼
Half can bottles	30.37 ½
Peppermint mistelle	6.37 ½
Cinnamon mistelle	3.00
Clove mistelle	3.00
Chinese mistelle	3.75
Barrel lids	1.50
Sweet wine	49.50
White wine	20.00
Compound wine	9.80
Nance aguardiente	0.62 ½

Source: AGEY, Fondo Justicia, Serie Civil, Subserie Testamentos Roque Milan, vol. 86, Expediente 25. Inventory and balance of a commercial house owned by D. Roque Milán and D. Esteban Martínez, situated on the corner called "El Elefante" (The Elephant), Mérida, Yucatán.

At this time aniseed was generically applied to all aguardientes obtained by distilling grounded anis, or *badiana*, with or without additional aromatic substances; in this case using cane alcohol, which could be supplemented with sugar, and reaching an alcoholic gradation higher than 30 percent. Based on their sugar content, three types of aniseed liquors were recognized: dry aniseed, with a sugar content between 0 and 20 g/L; half-dry aniseed, with a sugar content of 20–100 g/L; and sweet aniseed, with contents above 100 g/L. The latter was produced, traded, and consumed over the Yucatán Peninsula, with ground anise (a national import) used for maceration. However, additional aguardiente was purchased from other Mexican regions, including Tabasco, Mexico City, and Veracruz, while so-called "habanero rums" were imported from the Caribbean.

People in indigenous communities and hacienda workers, who were assigned a daily ration, consumed almost exclusively local aguardiente. The pervasiveness of this situation is best expressed by the words of a Peto town trader who commented on the scarcity of laborers in 1839: "Illnesses suffered by indigenous people could only be ameliorated by the use of aguardiente. The use of liquor in Yucatán, especially by indigenous people, contributes to their health, strength, and to the prevention of accidents. This has been proved extensively by lots of evidence"[9] (Lunuza 1839; my translation). Ranch and hacienda owners acquired the aguardiente destined for their workers, and the size of their orders depended on the size of their estate and the number of peasants who were employed. For example, the owners of hacienda *Oncan* declared in 1900 that "the estate had been provided with aguardiente for the servants for more than 20 years at the establishment called 'El Elefante'; the monthly consumption of two classes of aguardiente, *anisado* and *anisadillo*, reaches five hundred and fifty liters."[10] Given the volumes consumed by the indigenous population, the appropriate vessels to store, distribute, and commercialize them were casks, bottles, and flasks.

In 1902, the *City of Mérida's Classified Directory for Business Agencies and Mercantile and Industrial Establishments* advertised shops that offered a large variety of local aguardientes, as well as liquors and imported wines. Of course, the latter were destined for the consumption of a specific social sector: the elites. For example, Bolio and Marrufo, commissioned traders established at house number 175 on 65th Street, constantly sold, among other goods, superior rums from the state of Tabasco and homemade varieties, bitter anise from the south and the east of the state, the rich honey anise called *Xtabentun*, half or full barrels of sherry and red wines, bottled Macon wine, and Pauillac wine in medium bottles for the table.[11]

The consumption of alcoholic beverages at hacienda San Pedro Cholul

The hacienda San Pedro Cholul, located to the northeast of the city of Mérida, was a plantation devoted to the exploitation of henequen, and was inhabited for about one hundred years—between the second half of the nineteenth century and 1960. There has been archaeological research at the site over the past six years. The work consisted in exploring and excavating three household units locally known as *solares*, the hacienda's chapel, and several buildings belonging to its productive complex (main house, machinery room, warehouse, and administration). Since this hacienda is barely mentioned in written sources, our main objective has been to document the daily life of the plantation's peons and their subsistence strategies (Hernández Álvarez and Zimmermann 2016).

At San Pedro Cholul, the material evidence regarding the consumption of alcoholic beverages comes from a large amount of fragmented and whole glass bottles that were salvaged throughout the *solares* (patios). Glass was among the most abundant materials gathered both during surface collections and excavations (Hernández Álvarez, Venegas de la Torre and Solís Jarquín 2016). For example, within *Solar* 1 we collected 771 fragments, adding up to almost 11 kg. *Solar* 15 held the highest amount of glass, totaling 2,668 pieces and 35.5 kg (Hernández Álvarez and Escalante Kú 2012), while *Solar* 30 provided a further 2,281 shards, weighing a total of 27.75 kg (Hernández Álvarez and García Uitz 2014). The collection of complete vessels and bottles, which amounts to 204 specimens, gives us an idea of the different types of alcohol consumed in haciendas and also about patterns of local social distinction stemming from the different usage between landlords and workers. Beer, whisky, liquors and tonics, sparkling and red wines, cider, and champagne were among the alcoholic beverages that met the estate owner's taste. All of these drinks were important for, and highly valued by, the Gilded Age's Yucatecan elites.

Among the bottles found in San Pedro's *solares* there was at least one semi-complete bottle made from mold-blown glass. It is rectangular, colored green, and corresponds to the then famous tonic called *Aromatic Schnapps* by Udolpho Wolfe, dating between 1850 and 1920. This bottle displays the engraving "UDOLPHO WOLFE'S/SCHIEDAM/AROMATIC SCHNAPPS" on three of its four sides. The letter "Z" is set on its square base and might represent the glazier's brand. The glass is thick and one can observe air bubbles and a slight deformation. Both are distinctive characteristics of mold-blown bottles. In addition to these features, there is another clear indicator of the bottle's antiquity: it was plugged with cork. In 2010, three additional bottles of the same type were collected in the *Solar* 15. One was missing only the neck, the second was half-complete, and the third almost complete but totally fragmented (Hernández Álvarez and Escalante Kú 2012). This product's bottles have been found as part of the "Bertrand" cargo in the United States (Switzer 1974). In Yucatán, other finds of "Aromatic Schnapps" have been reported at the haciendas Pacabtún, Tabi, and San Antonio Xocnaceh (Benavides 1985; Burgos Villanueva, Palomo Carrillo and Dzúl Góngora 2005); shards have also been found at hacienda Anicabil, on the western edge of modern-day Mérida (Ligorred Perramón 2008).

Another of San Pedro's more interesting containers corresponds to a bottle of *Gancia* sparkling wine, which came from the Canelli region of Italy (Hernández Álvarez and Escalante Kú 2012). The bottle is olive-green in color, was mold-blown, and the neck shows an applied seal that reads "GANCIA & C./CANELLI" around the brand's escutcheon. Gancia sparkling wine was characterized by its sparkling bubbles and its dry taste; due to its simplicity, it was favored as an informal aperitif. The drink color was straw yellow with greenish flashes; it was described as having a soft, rich, and persistent sparkle, and a clear transparency. Its aroma evoked a mix of fresh fruits and an intense and consistent sweetness. It was considered to be the ideal wine to enjoy the pleasurable moments of life.[12]

Regarding beers, we have found remains from bottles of different brands and periods. For example, we tabulated a large amount of green- and amber-colored shards that correspond to vessels used to hold imported malts at the end of the nineteenth and beginning of the twentieth century. Dark shaded bottles were considered ideal for this type of beer, as they provide better protection from sunlight (Ortiz Castro 2009: 40). We also found several fragments from commercial beer bottles that were sold in Mexico since the beginning of the twentieth century. For example, there are amber-colored shards from the famous *Superior* brand, produced by Cervecería Moctezuma S.A. since 1902, flasks used for *Carta Clara* beers—brews locally produced by Cervecería Yucateca from 1900 until 1970—as well as Cuauhtémoc (Monterrey) products.

As we have demonstrated, the vast majority of bottles and glass shards recovered in the hacienda's *solares* were from wines, liquors, tonics, and beers that were consumed at San Pedro throughout the end of the nineteenth and beginning of the twentieth century. We suggest that the different documented alcoholic beverages correspond to the products consumed by the estate's owners, their families, and probably their guests. However, the presence of these drinks' material remains within the workers' *solares*, where they were not supposed to be found, shows that the bottles went through different processes of discarding, abandonment, reclaim, and reutilization during the site's different occupational periods (Hernández Álvarez 2015). For example, we noted previously that a significant quantity of bottles in excellent condition were recovered in *Solar* 15 (Hernández Álvarez, Venegas de la Torre and Solís Jarquín 2016). Descendants from members of this community informed us that those vessels were refilled with aguardiente and sold on-site to the hacienda's workers (Hernández Álvarez and Martín Medina 2016).

Concluding remarks

As we pointed out at the beginning of this chapter, so far the anthropological study of alcohol consumption has focused on different aspects related to cultural, economic, and health issues. However, we believe that one of the more interesting areas of current research is the role that the consumption of alcoholic beverages plays as an exclusionary mechanism among social classes. It has been historically documented that aguardiente production was one of the main economic activities on the Yucatán Peninsula during the second half of the nineteenth century. This production served to satisfy the increasing local market, which was consolidated through the consumption of cane liquors by indigenous people living on the region's haciendas. Meanwhile, the demand of alcoholic beverages by the peninsula's higher social classes was met through imports, both national and international, of a large variety of wines, liquors, and beers. The acquired products were traded at the principal businesses in Mérida, the state's capital.

Archaeologically, alcohol consumption has been documented through glass shards and bottles recovered from residential units at henequen haciendas such as San Pedro Cholul. This form of material culture is fundamental in demonstrating that differences in alcohol consumption patterns existed during the so-called golden age. We now know that the estate's owners enjoyed French and Italian wines, Dutch tonics, and English and US-American beers, while their workers consumed aguardiente rationed by the owners or bought at the hacienda's own store. Drinking is said to be an act of incorporation, due to the circulation of substances through the body. Commonly, this activity takes place in a social arena where the bodily senses are of fundamental importance. Hence the archaeology of the sensorium considers that bodies, material culture, and food and drink substances could encourage a sense of emplacement, intoxication, and altered states of consciousness under certain circumstances

(Hamilakis 2015; Sutton 2010). In this regard, the sensorial effects of drinking among the elite and the working class in Yucatec haciendas accentuated and enhanced distinct experiences and interactions that were socially marked.

Through the archaeological findings of large numbers of bottles that contained alcoholic beverages, including sparkling ones, we came to understand that peons did not consume certain goods (imported liquors, beers, and wines) but rather recycled and conserved objects (bottles), as the literature on poor people's consumption patterns suggests (Staski and Wilk 1985). This pattern of consumption stood in contrast to the landowners, whose consumption included several imported wines and liquors. Henequen workers recycled the landowners' discarded glass bottles to refill them with locally produced, cheap aguardiente. To conclude, we argue that a combined historical and archaeological perspective on aguardiente consumption allows us to better understand the social, economic, and political implications of alcohol usage among the Yucatán Peninsula's native populations, as well as its role as a mechanism of distinction between social classes in postcolonial times.

Notes

1 Registro Público de la Propiedad [Public Registry of Property] (1896), Archivo General del Estado de Yucatán [General Archive of the State of Yucatán] (AGEY), Book 1968.

2 While most national liquors came from the states of Tabasco, Veracruz, and Mexico, Saint Remy was a brandy of French origin.

3 Varieties of wine included classes such as dry, sweet, muscatel, sherry, and port, while liquors included gin, brandy, and cider.

4 In April 1896: 100 sacks weighing 2,150 kg.

5 In July 1896: 10 sacks weighing 3,906 kg; and in September: 1 sack weighing 57 kg.

6 In September 1896: 9 sacks weighing 129 kg.

7 In August 1896: 280 kg.

8 Fondo Poder Ejecutivo, Sección Ayuntamientos. Serie Correspondencia [Executive Power Fund, City Councils Section Correspondence Series] (1888), AGEY, Box 253.

9 "[L]as enfermedades que sufren los indígenas que quizá solo con el uso exterior del aguardiente estarían muchos buenos y sanos, puesto que el uso del licor en Yucatán, con especialidad en los indígenas, está comprobado hasta el último grado de evidencia, que contribuye muy mucho a la perfección de su salud, al reparo de sus fuerza, y a la preservación de muchos accidentes."

10 Fondo Justicia. Sección Juzgado Tercero de lo Criminal. Serie Penal. Subserie Defunciones. Asunto: la muerte de varios sirvientes de Oncaan, por intoxicación con aguardiente. Mérida [Justice Fund. Section Third Criminal Court. Criminal Series. Subseries Deaths. Subject: the death of several servants of Oncaan, due to intoxication with brandy. Mérida] (December 23, 1900–January 9, 1901), AGEY, vol. 160, file 13.

11 *Biblioteca Yucatánense. Guía Enciclopédica Popular. Directorio Clasificado de Agencias de negocios y establecimientos mercantiles e industriales de la Ciudad de Mérida* [Yucatecan Library Popular Encyclopedic Guide. City of Mérida's Classified Directory for Business Agencies and Mercantile and Industrial Establishments] (1902).

12 See: Casa Gancia, www.gancia.it (accessed December 10, 2017).

References

Ayora-Diaz, S. I. and G. Vargas-Cetina (2005), "Romantic Moods: Food, Beer, Music and the Yucatecan Soul," in T. Wilson, ed., *Drinking Cultures: Alcohol and Identity*, 155–178, Oxford: Berg.

Barrows, S. and R. Room, eds. (1991), *Drinking: Behavior and Belief in Modern History*, Berkeley: University of California Press.

Benavides, A. (1985), "Notas sobre la arqueología histórica de la hacienda Tabi, Yucatán," *Revista Mexicana de Estudios Antropológicos*, 31: 43–58.

Bittman, B. and G. Alcaide (1984), "Historical Archaeology in Abandoned Nitrate 'Oficinas' in Northern Chile: A Preliminary Report," *Historical Archaeology*, 18 (1): 52–75.

Bruman, H. (2000), *Alcohol in Ancient Mexico*, Salt Lake City: University of Utah Press.

Burgos Villanueva, R., Y. Palomo Carrillo and S. Dzúl Góngora (2005), *San Agustín de Pacabtún. arqueología e historia de una hacienda henequenera*, Mérida: Grupo Bepensa, Centro INAH-Yucatán.

Cámara Gutiérrez, G. (1995), "Destilación y comercio de aguardiente en Yucatán, 1821–1870," MA Thesis, Facultad de Ciencias Antropológicas, Universidad Autónoma de Yucatán, Mérida.

Cámara Gutiérrez, G. (2007), "Política arancelaria para la destilación y el comercio del aguardiente en Yucatán, 1821–1870," in P. Zabala Aguirre, P. Miranda Ojeda and J. Serrano Catzim, eds., *Poder político y control social en Yucatán, siglos XVI–XIX*, 151–185, Mérida: Universidad Autónoma de Yucatán.

Cámara Gutiérrez, G. (2009), "Aguardiente y transgresión en Yucatán, 1875–1895," in P. Miranda Ojeda and P. Zabala Aguirre, eds., *Normas, transgresiones. Infracciones al orden en la sociedad yucateca*, 247–270, Mérida: Universidad Autónoma de Yucatán.

Cámara Gutiérrez, G. (2010), "Alambiques y bebedores de aguardiente," in G. Negroe Sierra and P. Miranda Ojeda, eds., *Nuestra historia con minúsculas*, 87–102, Mérida: Biblioteca Básica de Yucatán, Gobierno del Estado de Yucatán.

Carey, D. (2012), "Distilling Perceptions of Crime: Maya Moonshiners and the State, 1898–1944," in D. Carey, ed., *Distilling the Influence of Alcohol: Aguardiente in Guatemalan History*, 120–156, Gainesville: University Press of Florida.

Dietler, M. (2006), "Alcohol: Anthropological/Archaeological Perspectives," *Annual Review of Anthropology*, 35: 229–249.

Douglas, M., ed. (1987), *Constructive Drinking: Perspectives on Drink from Anthropology*, Cambridge: Cambridge University Press.

Eber, C. (2000), *Women and Alcohol in a Highland Maya Town: Water of Hope, Water of Sorrow*, Austin: University of Texas Press.

Eckholm, E. and J. Deetz (1971), "Wellfleet Tavern," *Natural History*, 80 (7): 49–56.

Fallaw, B. (2002), "Dry Law, Wet Politics: Drinking and Prohibition in Post-Revolutionary Yucatán, 1915–1935," *Latin American Research Review*, 37 (2): 37–64.

Fontana, B. (1968), "Bottles and History: The Case of Magdalena de Kino, Sonora, Mexico," *Historical Archaeology*, 36 (2): 45–55.

Hamilakis, Y. (2015), "Food as Sensory Experience," in K. Metheny and M. Beaudry, eds., *Archaeology of Food: An Encyclopedia*, 205–206, Lanham: Rowman and Littlefield.

Heath, D. (1976), "Anthropological Perspectives on Alcohol: An Historical Review," in M. Everett, J. Waddell and D. Heath, eds., *Cross-Cultural Approaches to the Study of Alcohol*, 41–102, The Hague: Mouton.

Heath, D. (1987a), "Anthropology and Alcohol Studies: Current Issues," *Annual Review of Anthropology*, 16: 99–120.

Heath, D. (1987b), "A Decade of Development in the Anthropological Study of Alcohol Use," in M. Douglas, ed., *Constructive Drinking: Perspectives on Drinking from Anthropology*, 16–69, Cambridge: Cambridge University Press.

Hernández Álvarez, H. (2015), "Capitalism and Material Culture of the Poor: Consumption, Reuse, and Discard of Glass Bottles at Hacienda San Pedro Cholul, Yucatán," paper presented at the 80th Annual Meeting of the Society for American Archaeology, San Francisco: Society for American Archaeology.

Hernández Álvarez, H. and G. Cámara Gutiérrez (2017), "El comercio y consumo de aguardiente en Yucatán hacia finales del siglo XIX y principios del siglo XX," in M. Del Rosario Domínguez Carrasco, M. J. Gallegos Gómora, R. Armijo Torres and M. E. León Méndez, eds., *Los Investigadores de la Cultura Maya. El comercio y otros temas*, 265–278, Campeche: Universidad Autónoma de Campeche.

Hernández Álvarez, H. and T. Escalante Kú (2012), "Vidrios y botellas," in H. Hernández Álvarez, L. Fernández Souza and M. Zimmermann, eds., *Proyecto Arqueología Histórica en la Hacienda San Pedro Cholul*, Informe Final, Temporada de Campo, septiembre 2009–mayo 2010, manuscript, 135–162, Mexico City: Consejo de Arqueología, Instituto Nacional de Antropología e Historia.

Hernández Álvarez, H. and Á. García Uitz (2014), "Análisis de los elementos de vidrios del Solar 30," in H. Hernández Álvarez, ed., *Proyecto Arqueología Histórica en la Hacienda San Pedro Cholul*, Informe Final, Temporada de Campo 2013, manuscript, 151–180, Mexico City: Consejo de Arqueología, Instituto Nacional de Antropología e Historia.

Hernández Álvarez, H. and G. Martín Medina (2016), "Arqueología colaborativa y recuperación de la memoria histórica. Hacienda San Pedro Cholul, Yucatán," *Temas Antropológicos*, 38 (2): 109–127.

Hernández Álvarez, H. and M. Zimmermann, eds. (2016), *Sendas del henequén. un estudio arqueológico de la hacienda San Pedro Cholul, Yucatán*, Mérida: Universidad Autónoma de Yucatán.

Hernández Álvarez, H., L. J. Venegas de la Torre, and R. Solís Jarquín (2016), "Patrones de consumo y desecho en dos grupos domésticos de principios del siglo XX," in H. Hernández Álvarez and M. Zimmermann, eds., *Sendas del henequén. Un estudio arqueológico de la hacienda San Pedro Cholul, Yucatán*, 143–170, Mérida: Universidad Autónoma de Yucatán.

Jennings, J. and B. Bowser, eds. (2008), *Drink, Power, and Society in the Andes*, Gainesville: University Press of Florida.

Karskens, G. (2003), "Revisiting the Worldview: The Archaeology of Convict Households in Sydney's Rock Neighborhood," *Historical Archaeology*, 37 (1): 34–55.

Kent, S. (1983), "The Differential Acceptance of Culture Change: An Archaeological Test Case," *Historical Archaeology*, 17 (2): 56–63.

Ligorred Perramón, J. (2008), "Proyecto arqueológico Hacienda Anikabil, municipio de Mérida," Mérida: Departamento de Patrimonio Arqueológico y Natural, Dirección de Desarrollo Urbano, Ayuntamiento de Mérida.

Lunuza, J. M. (1839), *Exposición comerciantes y labradores de Peto hacen al Exmo. Sr. Gobernador de este departmento sobre el gravoso impuesto que sufren las destilerías de aguardiente (folletería)*, Mérida: Imprenta de Lorenzo Seguí.

Miranda Ojeda, P. (2010), "Traficantes y bebedores de aguardiente," in G. Negroe Sierra and P. Miranda Ojeda, eds., *Nuestra historia con minúsculas*, 77–86, Mérida: Biblioteca Básica de Yucatán, Gobierno del Estado de Yucatán.

Noël Hume, I. (1968), "A Collection of Glass from Port Royal, Jamaica, with Some Observations on the Site, Its History, and Archaeology," *Historical Archaeology*, 2: 5–34.

Ortiz Castro, C. (2009), *Botellas de vidrio: bases para un catálogo arqueológico de Colombia*, Bogotá: Facultad de Ciencias Sociales-CESO, Departamento de Antropología, Universidad de los Andes.

Ortiz Yam, I. (2013), *De milperos a henequeneros en Yucatán, 1870–1937*, Mexico City: El Colegio de México.

Peniche Rivero, P. (2010), *La historia secreta de la hacienda henequenera de Yucatán. Deudas, migración y resistencia maya*, Mérida: Instituto de Cultura de Yucatán.

Schwartzkopf, S. (2012), "Consumption, Custom, and Control: Aguardiente in Nineteenth-Century Maya Guatemala," in D. Carey, ed., *Distilling the Influence of*

Alcohol: Aguardiente in Guatemalan History, 17–41, Gainesville: University Press of Florida.

Smith, F. (2008), *The Archaeology of Alcohol and Drinking*, Gainesville: University Press of Florida.

Staski, E. and R. Wilk (1985), "La cultura material de áreas marginales y gente pobre. un caso del distrito de Toledo, Belice," *Revista Mexicana de Estudios Antropológicos*, 31: 155–161.

Sutton, D. (2010), "Food and the Senses," *Annual Review of Anthropology*, 39: 209–223.

Switzer, R. (1974), *The Bertrand Bottles: A Study of 19th-Century Glass and Ceramic Containers*, Washington, DC: National Park Service, US Department of the Interior.

Torras Conangla, R. (2007), "La embriaguez del poder: los estancos de aguardiente y la penetración ladina en los pueblos mames guatemaltecos. El caso de Colotenango," *Península*, 2 (2): 97–119.

Wells, A. (1985), *Yucatán's Gilded Age: Haciendas, Henequen, and International Harvester 1860–1915*, Albuquerque: University of New Mexico Press.

Wells, A. and G. Joseph (2011), *Verano del descontento, épocas de trastorno: elites políticas e insurgencia rural en Yucatán, 1876–1915*, Mérida: Universidad Autónoma de Yucatán.

Wilson, T., ed. (2005), *Drinking Cultures: Alcohol and Identity*, Oxford: Berg.

4

The Flavors of Corn: A Unique Combination of Tradition and Nature

Mario Fernández-Zarza
Universidad de Sevilla

Ignacio López-Moreno
Universidad Autónoma Metropolitana

Introduction: Beyond taste in the mouth

In this chapter we focus on the description and analysis of the flavors of corn in Mexican gastronomy, taking the central Mexican highlands as our main focus. There are many ways to conceive and analyze flavor, and our contribution is informed by approaches that understand reality as a socioculturally constructed phenomenon (Ayora-Diaz 2012; Bourdieu [1984] 1998; Cárdenas-Carrión 2014; Contreras-Hernández 1995; Díaz Méndez and Gómez Benito 2005; Fischler 1995; Harris 1998; Levenstein 2012; Levi-Strauss 1968; Pollan 2014). This is a complex task and one that is difficult to pursue given that maize is considered a "super food,"[1] with more than 9,000 years of history and coevolution in the American continent. In this sense, it is important to remember that Mexico is considered to be the center of origin and domestication of this cereal (López-Moreno and Vizcarra-Bordi 2016; Sarukhán 2013) and one of the eight global nodes where agriculture began (Ortiz-García and Otero-Arnaiz 2007).

This cereal is the product of a coevolutionary process that dates back to the years between 8,990 and 8,610 BCE (González-Jácome 2016), but did not become a fundamental element of the diet until 1,000 BCE (Oseguera-Parra and Ortega-Paczka 2016). This long process was predicated on the social, ecological, and cultural transformations of all the elements encompassed by a system that is still developing today. The result of this process is the creation of more than sixty races and 300 varieties of native corn that have *teozintle* as a common ancestor, and with which more than 700 dishes are being prepared today within the Mexican republic (Dávalos Hurtado 2012). In this sense, it is important to affirm, from the beginning of this chapter, that societies construct a system of classification based on perceived flavors, and this classification is in turn a manifestation, and meaning, of a group's alimentary culture (Cárdenas-Carrión 2014). Each society perceives and reproduces its own sensorial construction; that is, every culture constructs and interprets meanings attributed to the sensations perceived when food is introduced into the mouth—or not necessarily introduced (Mazzetto 2017).

Brillat-Savarin ([1825] 1854) already proposed some two centuries ago that "gastronomy" is a discipline that encompasses both sensorial and contextual aspects not included previously: first, the physical aspect that refers to the simple act of appreciating and valuing flavor; second, the moral dimension, due to the impression it grafts onto the organism; and third, the material dimension, on account of the sensations that are created around food. Therefore, the perception of flavor is not a simple process that ends with chewing food; on the contrary, flavor is an essential part of human perception related to nourishment, as well as a cultural construct subject to transformations, reinterpretations, and social adaptations though different times and places.

Beginning with the notion that sensorial perceptions are linked to forms of cultural appropriation and belonging, it is then necessary to examine the relationship between: (1) sensory meanings; (2) the habitus generated by societies and their culture; (3) the traits that are implied; (4) the stimuli; (5) the appreciations/valuations; and (6) the experiences laden with collective memories (Cárdenas-Carrión 2014). To achieve this, the anthropology of the senses, as developed by Classen (1997), highlights that sensorial and organoleptic perceptions are cultural. In other words, the five senses are not exclusively learnt through physical reactions to the world but, rather, they are supplemented by shared cultural values. Therefore, it is not only through tongue sensations that we perceive sweet, salty, sour, acid, and *umami* flavors, but also important for the experience of flavor is the synesthetic sensorial experience that involves the perception of temperatures, textures, sounds, caresses, scents, and colors (Cárdenas-Carrión 2014). Hence, the study of alimentary culture allows us to understand how "good flavor" is classified and deemed acceptable within a society (Johnston and Baumann 2014). Perception through the senses is, in fact, regulated by society and carries with it cultural impositions and obliterations that shape the way in which human beings categorize what is good, tasty, delicious, or nutritious (Ayora-Díaz 2014; Harris 1998). Thus, this is the process where flavor communicates forms and meanings into sensorial models dominant in a society, shaping ideas of what is better for the world, more deserving of awards, more "authentic," "purer," and what ultimately is important here: food marks and is marked by social status (López-Moreno and Aguilar Criado 2012; Thomé-Ortiz and Moctezuma-Pérez 2015).

Flavor, which merges with smell, visual, and touch sensations, represents much more than a straightforward organoleptic sensory impression or response. The articulation of the senses is a social construct tied to multiple, coexisting time lines. Nonetheless, we need to acknowledge two other coexisting phenomena: first, there are diverse flavors distinguishable from, or resulting from, multiple amalgamations of salty, sweet, acid, and *umami* perceptions. Second, in the last two decades, the codes emerging from what has been called *liquid globalization* suggest a tendency to impose homogenizing standards for flavors; standards that would be available to any person and which could be reproduced in any place at any time. This is further complicated by existing flavor hierarchies that are tied to the subjects belonging to a given social class, and the power conferred to, and drawn from, aesthetic preferences used to distinguish between social groups (Johnston and Baumann 2014).

For example, at the beginning of the twentieth century, Mexican elites launched themselves against corn-based diets, since they were perceived as an obstacle for "Progress" (and was later used to explain "underdevelopment" in Mexico). Complementarily, scientific discourse, particularly the one produced by nutrition specialists, argued that corn was inferior to wheat and, therefore, Indians—indigenous people—would have to be taught to eat European grains. Pilcher ([1998] 2001) has called this shift of meanings "*tortilla* discourse" where at the end of the nineteenth century corn and tortillas were despised by politicians with the aid of scientific discourse, and later on, during the first half of the twentieth century, with the Mexican revolution, corn and tortillas became central to the fabric of Mexican nationalism and identity. Despite the former discourse, agricultural and industrial modernization was unable to replace *tortillas* but succeeded in changing it into a commodity, transforming corn from a subsistence product into a market good (Pilcher [1998] 2001). With the passing of time this discourse grew stronger in the urban context while in Mexican rural settings, probably due to their isolation, the negligible presence of a currency-driven economy, and to the cultural force of diverse cultural identities, the production, storage, transformation, and cooking of food became discursively tied to the indigenous past (Arroyo 2006). At the end of the twentieth century and beginning of the twenty-first this understanding has been turned upside down: corn is today a national symbol that stands at the center of the discourse on *Mexicanidad* (Mexican identity), while agri-food corporations marginalize and reduce the biodiversity of the *milpa* (maize field), and promote cuisines based on alternative cereals.

Notwithstanding changes in discourses and perceptions, maize remains the basis of flavor analysis that we discuss below. The importance of corn/maize as a superfood is comparable to rice in Asia and wheat in Europe. This is why it is important to look at and examine the basic foods consumed within each culture as indicators of their culinary complexity and of the presence of culturally meaningful alimentary systems. We can thus recognize which foods are peripheral and occasionally consumed, and distinguish them from the superfoods consumed on a daily basis. Hence, we need to analyze the relationship between a superfood and its multiple gastronomical constructions and confections, as indicators of their importance within an alimentary culture—maize in the Mexican case. This can help us understand how, despite growing public disaffection with the place of corn in their everyday diet, it has survived in the taste of Mexican society at large. Furthermore, currently we are witnessing the reemergence of a large variety of corn with diverse flavors, colors, and textures.

However, in general, the diversity of maize has considerably declined, among other things, due to the fact that farmers have been forced to seek a higher income to support their families and, hence, are adopting commercial varieties of homogeneous corn. This has an impact on both the biological diversity of maize and on the diversity of culinary preparations that are achieved through the use of specific races and varieties. Ultimately, this practice promotes a significant change on the diet of societies, particularly among rural ones (Moreno-Flores et al. 2014).

A constant effort to homogenize flavors

Before analyzing the culture of maize it is important to discuss the current alimentary conundrum affecting the construction and reproduction of flavors within society. We believe this is important as we recognize that cultural values shape the collective construction of what "food" is, as well as the food habits of any society: you are what you eat and where you come from (Belasco 2002). Hence, in writing about alimentary culture we recognize the importance of practices, representations, imaginaries, and classifications, as they reveal different types of food preferences each related to a specific gastronomic context. Culinary norms connect preparations, textures, and flavors, giving social meaning to everyday acts of eating. Thus, the growth in the number of studies examining the relationship between culture and food has helped develop new and refreshing theories and methodologies to conduct our analysis. This field brings together theoretical, methodological, and conceptual frameworks from different disciplines that enhance sociological and anthropological analysis (Aguilar 2014; Díaz-Mendez and Gómez-Benito 2005). Thanks to the development of food and nutrition studies, we are now able to identify cultural differences arising from the relationship between territory and gastronomy. Hence, we have tools to explain how some members of any society privilege the enjoyment of food, while others become technologically-disposed and choose their food for its nutritional rather than its pleasurable value.

Gastronomies such as the ones from Mexico, Spain, Italy, Colombia, Nicaragua, Senegal, Zimbabwe, Taiwan, or New Zealand, to name but a few, are based on ingredients, processes, and forms of consumption that distinguish them from each other, even when they share a large number of common elements. Their differences suggest that when people become acquainted and interact with unfamiliar gastronomies, they tend to classify them as "exotic" foods, introducing ambivalences proper to the distinction between "exotic" and "authentic" foods. Nonetheless, these two terms go hand in hand, as an "exotic" food can be "authentic" at the same time depending on the context (Johnston and Baumann 2014). From this perspective, global processes shape the convergence of producers' and consumers' practices where their food practices are the result of a cultural negotiation between local and foreign flavors (James 2005).

This ambivalence is exploited by a food industry eager to homogenize preparations in order to gain a larger share in the alimentary market. Gradually, consumers are adopting food habits dependent on the use of processed and hyperprocessed foods, and is evidenced by the growing number of people who find artificially produced flavors more pleasant than "natural" ones. Here we face a process of *naturalization of taste* where large global agricultural and food industries encourage the adoption of their products among the Mexican population (Ayora-Díaz 2017). Consequently, industrially processed fats that are more easily distributed through the market replace natural lipids. In addition, there is a higher consumption of high-energy products (sweets, sodas, and industrial breads) enhanced with chemically produced flavors that appeal to people of all ages and genders (Contreras-Hernández and Ribas-Sierra 2012). Furthermore, these products are underwritten by science: the most "reliable" source of information in this era when scientific[2] practices deployed in laboratories transfer and validate the belief of nutritional virtues ascribed to the labels of commercial food products (Nestle 2013). Thus, societies witness the decline of culinary knowledge received from their ancestors, as well as of other practices tied to the enjoyment of foods, such as that derived from food textures, food-centered sociality, and even eating with one's hands (Ayora-Díaz 2014). This differs from what has been understood as the historical knowledge of foods, which is acquired through time with the help of the senses; a knowledge that is now replaced by alimentary science—product of hegemonic cultures—tasked with defining what the "right" food is. This displacement of local knowledge has encouraged the abandonment of traditional diets and the search for "healthier" food alternatives effective to control the body (Contreras-Hernández and Ribas-Serra 2012).

Today, the ideology of distinction and social status supports the view that only certain privileged individuals can appreciate and understand "the correct way to eat" (Johnston and Baumann 2014). For example, in Mexico there is a tendency to adopt many of the North Atlantic preoccupations, diets, and theories, particularly European ones that value wheat over corn to define what it means to "eat well" (Benito-Morán 2016; Contreras-Hernández 1999). It is clear that food recommendations that seek to promote the consumption of certain products while eliminating others have several possible consequences, such as undermining traditional knowledge and elevating technological, specialized knowledge, leading to *alimentary disaffection* (Gliessman 2014).

Alimentary disaffection is further complicated by gender subordination: while food is meaningful when it refers to male domains (such as production and selling), actual cooking in everyday life is seen as restricted to the private female sphere, and is granted less social importance (Johnston and Baumann 2014). In regard to this distinction, Ayora-Díaz (2017) emphasizes a decisive aspect for the creation of flavors that mirrors Bourdieu's concept of *hexis*, dealing with public and private spheres. Flavors highly desired by local people are created in the private sphere, one reserved almost exclusively for housewives. In contrast, restaurant and foreign eating establishments are the loci of choice in the public sphere, where local foods are reappropriated. This is illustrated by current gastronomic fashions such as the restaurants Pujol, or by the brief opening of NOMA in the Riviera Maya, where local, rural dishes and flavors were served at exorbitant prices (Wells 2017).

Peasants' food is usually disavowed by the higher social classes, who would prefer European high cuisine instead of sharing tamales with the "people down below." Indigenous cuisine triggers a dichotomous opposition: on one hand, it invokes a foundational identity and awakes nostalgic feelings. On the other, it provokes disdain and rejection on account of its devalued ethnic source (Pilcher [1998] 2001). This ambivalence is grounded on three social phenomena: the romantic understanding of poor people's food when it is seen as the "true" food of the folk; the spectacularization of inequality, such as when the food of "poor people" is displayed in the "gastronomic" menus of fashionable expensive restaurants; and the normalization of privileges, which turns these inequalities into everyday "reality" (Johnston and Baumann 2014). All societies, including Mexico's, have been divided since pre-Hispanic times between the cuisine of those who have less, and the cuisine of the ruling parties and the elites (Ayora-Díaz 2017). In contrast, the nonobjectified cuisines, such as the maize's cuisine, possess a greater nutritional value, and it constitutes the culinary practices of subordinated groups within society (Flandrin 1999; Goody 2006; Trubek 2000).

Maize's flavors, Mexican flavors

In all countries people construct their identity selecting, classifying, and hierarchizing symbolic and material aspects of their history and the territory they live in. From a distance, Mexican cuisine may be seen as homogeneous but *in situ* it is diverse, and it tastes differently according to the region. The historic record shows the evolution and geographic distribution of maize in Mexico and in the Americas as a foundational food (Cheetham 2010; González-Jácome 2016). Culinary practice shows that maize has various properties and traits: it can be the main ingredient or the meal itself; or it can be the substance to enfold food or a thickener; it can be a "container," sealer, or a fermenting ingredient; this versatility has led to maize being understood as a cultural superfood. When so many different techniques are applied to maize—such as *nixtamalization* (soaking in limewater), boiling, steaming, whisking, aeration, *tatemado* (toasting), grilling, fermenting, *pib* (baked underground), or frying—many flavors emerge, including sweet, salty, *umami*, sour, and acidic. Moreover, maize's culinary potential is so vast that when it is harvested at different stages of growth it is used to cook dishes for which different flavors and properties are expected and recognized (Centurión-Hidalgo et al. 2016). A paradigmatic example of this diversity of flavors is *pozol* (Green 2010), a ritual and energizing beverage resulting from the fermentation of maize. It has a sour taste stemming from its two or three day-long process of fermentation. It is often drunk adding sugar to sweeten it. However, some add chili pepper, chocolate, or other aromatic ingredients, giving it a different flavor, aroma, and taste. The flavors in Mexican food are usually strong and closely related to the recognition of the people with whom meals are shared and consumed. Additionally, central Mexicans find moisture in their food a pleasant attribute, and for Mazzetto (2014), for example, this explains the common use of *salsas* in meals in different Mexican regions.

Outside Mexico it is often assumed that people consume heavily hot-spiced foods that erase flavors. However, although it is true that there persists a high consumption of chili peppers in most of Mexico, and that chilies may conceal certain flavors, the throbbing (hot) sensation can be less intense than imagined, especially when the person has a long-lasting habit of eating them. It is certainly true that in several Mexican regions food is almost always accompanied by salsa that usually contains chili. Still, what cooks seek is to enhance the flavor of some meals. Although Mexico is far from being the country with the highest consumption of hot spices (Ayora-Díaz 2017), some foreign consumers perceive its flavor as marking Mexican food as "exotic" and "authentic."

In Mexico, in maize-based meals flavors can be as diverse as the ways in which they are prepared. All of the previously mentioned ingredients are combined and blended to please the Mexican palate, so that at the same time, and in a single meal, one may enjoy sweet and salty flavors. This is the case with pork rib tamales in Veracruz, which are prepared with tender maize, spicy *chipotle* salsa, pork fat, sugar, salt, and *hoja santa* (*piper auritum*). Sweet, salty, and spicy flavors bring different sensations into the consumption of the meal. For that experience, in ancient times inhabitants of the central Mexican highlands used the term *atoyaxocotl* to make reference to the simultaneous sensation of sweet and sour (Mazzetto 2017). In an experiment that Mazzetto (2017) performed with a group of *nahuas* (an indigenous group in central Mexico), he found that indigenous people were able to identify sweet, mellow, bitter, acid, tasty, smelly, thick, watery, warm, cold, sour, spicy, and salty flavors. Also, the Maya people of Yucatán resort to corn fermentation and acidification before shaping small balls that they carry on journeys that can last many months. However, the construction of flavor depends on the complementary work of interpretation and construction of meaning that people conduct. Soft, crude, porous, warm, and crisp, are all adjectives used to assign sensorial qualities to food. Flavors, textures, and odors contribute to satisfy even the strictest consumers.

The flavor resulting from the mixture of sweet and fatty produces great pleasure among many people. One can find different sweet flavors pleasing to the palate among maize-based meals. Today refined sugar has become the substitute for natural flavors, and the Mexican palate has adapted itself to seek a stronger candied flavor, leading to forms of *alimentary disaffection* where the flavors of the maize are gradually devalued. Maize has a natural sugar, perceived as sweetness in the flavor of tortillas, breads, and other tender maize-based preparations. Also, while in North Atlantic societies acid and sour flavors are often rejected, in Mexico there is great affection for these flavors. The use of lime, lemon, and other acidic elements highlight a particular flavor in dishes that when combined with other elements such as chili, onion, and coriander, demonstrate an additional characteristic of corn-based gastronomy. Thus, one can observe the use of lime juice in tacos, *pozole, ceviches* and *aguachiles* (which are served on corn tostadas), *esquites* (toasted corn), corn husks, and lime soup.

In contrast, while in some cultures people reject sour flavors, in the case of Mexico it is (or was) a common flavor in different meals. Such is the case of the fungus *huitlacoche* (*Ustilago maydis*), a corn byproduct highly valued in central Mexico that in other cultures would not be so on account of its black aspect and sour flavor. This happens as well with some bitter or oddly smelling herbs such as *epazote*, avocado, and *chaya* leafs. These almost always mark maize-based dishes as indexical of "poverty." Hence, when examining the flavors present in tables of the "poor," these could be described as a mixture of acid, sour, sweet, salty, and *umami* (earthy) flavors, which add complexity to the different flavors that different races and varieties of corn are accorded in Mexican culture. In addition, we need to remember that these are meals consumed in social and family events, providing maize with the potential to awake sensations and connect them to affective meanings.

Conclusions

In this chapter we have discussed the complexity of flavors that mostly central Mexicans recognize in corn, a superfood that has been part of Mexican prehistory and history for the last 10,000 years. This text belongs within a tradition of critical studies in the social sciences that understand flavor as a sociocultural construction. Following this approach, our work shows the importance that both social learning and social groups play in the appreciation and interpretation of flavors. If it is true that corn is the fundamental nutritious element in Mexican cuisine, our work reveals the weak potential of contemporary corn strains to recreate a wide spectrum of flavors. This is because flavors depend on corn's race, variety, and development; its mode of preparation; the recipes employed; the form of consumption; and the cultural significance it is assigned. The development of agricultural and alimentary industries during the last decades, alongside public policies that have promoted the abandonment of agricultural fields and encouraged the importation of hybrid and transgenic maize, have put the diversity of maize's flavors at risk. In sum, corn's flavors are a social construction located in the transition from Mexican to international gastronomies. The diversity of techniques for preparation and modalities of consumption are the offspring of a process of coevolution that has generated hundreds, if not thousands, of distinct agro-systems throughout human history. The combination of this cereal with other ingredients and within different traditions continues to multiply recipes and preparation techniques, resulting in a diversity of dishes and flavors throughout the world. These dishes and flavors depend on the production and reproduction of social and biological elements, in culture and nature, which are now affected by both climate change and the homogenizing effects of industrial mass-produced ingredients on different gastronomic traditions. Corn has played an important part in shaping Mexican culinary practices, flavors, and identities. We suggest we have to be aware of contemporary changes in the latter, as powerful global forces are impinging on local and translocal processes.

Notes

1 Jelliffe (1967) defined cultural superfoods as all of those that fulfill a central role in both the caloric intake of a society's diet and, in addition, possess a symbolic value.

2 Evidently, this argument reflects what Luhmann (1998) suggested when he affirmed that the "scientist" does not possess the absolute truth but rather one of several possible explanations.

References

Aguilar, P. (2014), "Cultura y alimentación. Aspectos fundamentales para una visión comprensiva de la alimentación humana," *Anales de Antropología*, 48 (1): 11–31.

Arroyo, P. (2006), "Estudios sobre Antropología y Nutrición en México," in M. Bertran and P. Arroyo, eds., *Antropología y Nutrición*, 13–25, Mexico City: Universidad Autónoma Metropolitana.

Ayora-Díaz, S. I. (2012), *Foodscapes, Foodfields and Identities in Yucatán*, Amsterdam: CEDLA, New York: Berghahn.

Ayora-Díaz, S. I. (2014), "El performance de lo yucateco: cocina, tecnología y gusto," *Alteridades*, 24 (48): 59–69.

Ayora-Díaz, S. I. (2017), "Translocalidad, globalización y regionalismo. Cómo entender la gastronomía regional yucateca," *Anales de Antropología*, 51 (2): 96–105.

Belasco, W. (2002), *Food Matters: Perspectives on an Emerging Field*, New York: Routledge.

Benito-Morán, C. de (2016), *La alimentación como práctica política cotidiana. Análisis antropológico social de los grupos de consumo ecológico*, Madrid: Universidad Autónoma de Madrid.

Bourdieu, P. ([1984] 1998), *La distinción. Criterios y bases sociales del gusto*, Barcelona: Taurus.

Brillat-Savarin, J. A. ([1825] 1854), *The Physiology of Taste: Transcendental Gastronomy*, Philadelphia: Lindsay & Blakiston.

Cárdenas-Carrión, B. M. (2014), "Construcciones culturales del sabor. Comida rarámuri," *Anales de Antropología*, 48 (1): 33–57.

Centurión-Hidalgo, D., J. Espinosa-Moreno, M. D. Reyes-Duarte, M. C. Wacher-Rodarte and G. Díaz-Ruiz (2016), "Alimentos tradiciones que se generan en la cocina rural tabasqueña durante el desarrollo de la mazorca de maíz," in I. López-Moreno and I. Vizcarra-Bordi, eds., *El maíz nativo en México. Una aproximación desde los estudios rurales*, 137–172, Mexico City: UAM-Juan Pablos editores.

Cheetham, D. (2010), "Corn, Colanders, and Cooking: Early Maize Processing in the Maya Lowlands and Its Implications," in J. Staller and M. Carrasco, eds., *Pre-Columbian Foodways*, 345–368, New York: Springer.

Classen, C. (1997), "Foundations for an Anthropology of the Senses," *International Social Science Journal*, 49 (153): 401–412.

Contreras-Hernández, J. (1995), *Alimentación y cultura. Necesidades, gustos y costumbres*, Barcelona: Universitat de Barcelona.

Contreras-Hernández, J. (1999), "Cambios sociales y cambios en los comportamientos alimentarios en la España de la segunda mitad del siglo XX," *Anuario de Psicología/ The UB Journal of Psychology*, 30 (2): 25–42.

Contreras-Hernández, J. and J. Ribas-Serra (2012), "Los alimentos modificados. El omnívoro desculturizado," *Gazeta de Antropología*, 28 (3), art. 04.

Dávalos Hurtado, Eusebio (2012), *Alimentos básicos e inventiva culinaria del mexicano*, México: SEP.

Díaz Méndez, C. and C. Gómez-Benito (2005), "Sociología y alimentación," *Revista Internacional de Sociología*, Tercera Época, 63 (40): 21–46.

Flandrin, J.-L. (1999), "From Dietetics to Gastronomy: The Liberation of the Gourmet," in J.-L. Flandrin and M. Montanari, eds., *Food: A Culinary History*, 418–432, New York: Columbia University Press.

Fischler, C. (1995), *El (h)omnívoro. El gusto, la cocina y el cuerpo*, Barcelona: Anagrama.

Gliessman, S. (2014), *Agroecology: The Ecology of Sustainable Food Systems*, New York: CRC Press, Taylor and Francis Group.

González-Jácome, A. (2016), "Orígenes, domesticación y dispersión del maíz (zea mays) en México," in I. López-Moreno and I. Vizcarra-Bordi, eds., *El maíz nativo en México. Una aproximación crítica desde los estudios rurales*, 25–63, Mexico City: Universidad Autónoma Metropolitana-Juan Pablos Editor.

Goody, J. (2006), "Gordon Childe, the Urban Revolution, and the Haute Cuisine: An Anthropo-Archaeological View of Modern History," *Comparative Studies in Society and History*, 48 (3): 503–519.

Green, J. S. (2010), "Feasting with Foam: Ceremonial Drinks of Cacao, Maize, and Pataxte Cacao," in J. Staller and M. Carrasco, eds., *Pre-Columbian Foodways: Interdisciplinary Approaches to Food, Culture, and Markets in Ancient Mesoamerica*, 315–343, New York: Springer.

Harris, M. (1998), *Good to Eat: Riddles of Food and Culture*, Long Grove, IL: Waveland Press.

James, A. (2005), "Identity and the Global Stew," in C. Korsmeyer, ed., *The Taste Culture Reader: Experiencing Food and Drink*, 372–384, New York: Berg.

Jelliffe, D. B. (1967), "Parallel Food Classifications in Developing and Industrialized Countries," *The American Journal of Clinical Nutrition*, 20 (3): 279–281.

Johnston, J. and S. Baumann (2014), *Foodies: Democracy and Distinction in the Gourmet Foodscape*, London: Routledge.

Lévi-Strauss, C. (1968), *Lo Crudo y lo Cocido. Mitológicas I*, Mexico City: Fondo de Cultura Económica.

Levenstein, H. (2012), *Fear of Food: A History of Why We Worry about What We Eat*, Chicago: The University of Chicago Press.

López-Moreno, I. and E. Aguilar Criado (2012), "Las etiquetas de calidad y el desarrollo territorial. Los casos del queso de oveja merina de Grazalema y la carne de cordero de Texel," *Revista de Economia Agrícola*, 59 (2): 131–149.

López-Moreno, I. and I. Vizcarra-Bordi, eds. (2016), *El maíz nativo en México. Una aproximación desde los estudios rurales*, Mexico City: Universidad Autónoma Metropolitana-Juan Pablos editores.

Luhmann, N. (1998), *Sistemas sociales. Lineamientos para una teoría general*, Mexico City: Anthropos Editorial, Universidad Iberoamericana and Centro Editorial Javierino.

Mazzetto, E. (2014), "Le metafore alimentari e vegetali del guerriero defunto nella cultura azteca," in A. Vargas, ed., *Studi Tanatologici*, 135–156, Torino: Fondazione Ariodante Fabretti ONLUS.

Mazzetto, E. (2017), "Un acercamiento al léxico del sabor entre los antiguos nahuas," *Anales de Antropología*, 51 (2): 154–170.

Moreno-Flores, M. E., I. Vizcarra-Bordi, A. D. Benítez-Arciniega and C. Chávez-Mejía, (2014), "El grupo de alimentos del maíz en la estructura energética de la dieta de madres de hogares productores de maíces nativos en dos comunidades del centro de México," *Revista Española de Nutrición Humana y Dietética*, 18 (2): 68–73.

Nestle, M. (2013), *Food Politics: How the Food Industry Influences Nutrition and Health*, Berkeley: University of California Press.

Ortiz-Garcia, S. and A. Otero-Arnaiz (2007), "México como el centro de origen del maíz y elementos sobre la distribución de parientes silvestres y variedades o razas de maíz en el norte de México," *Revista de Geografía Agrícola*, 38: 141–152.

Oseguera-Parra, D. and R. Ortega-Paczka (2016), "Gente de maíz. Historia y diversidad en la cocina mexicana del maíz," in I. López-Moreno and I. Vizcarra-Bordi, eds., *El maíz nativo en México. Una aproximación crítica desde los estudios rurales*, 113–136, Mexico City: Universidad Autónoma Metropolitana-Juan Pablos Editor.

Pilcher, J. ([1998] 2001), *¡Qué vivan los tamales! La comida y la construcción de la identidad nacional*, Mexico City: CIESAS and CONACULTA.

Pollan, M. (2014), *Cocinar. Una historia natural de la transformación*, Barcelona: Debate.

Sarukhán, J. (2013), "Prefacio," in E. R. Álvarez-Buylla and A. Piñeyro-Nelson, eds., *El maíz en peligro ante los transgénicos*, 11–15, Mexico City: Universidad Nacional Autónoma de México- Centro de Investigaciones Interdisciplinarias en Ciencias y Humanidades and Unión de Científicos Comprometidos con la Sociedad.

Thome-Ortiz, H. and S. Moctezuma-Pérez (2015), "Leisure Time and Food Memory: An Anthropological Approach to Culinary Tourism in Central Mexico," *Folia Turistica*, 37: 69–83.

Trubek, A. (2000), *Haute Cuisine: How the French Invented the Culinary Profession*, Philadelphia: University of Pennsylvania Press.

Wells, P. (2017), "Why I'm Not Reviewing Noma Mexico," *New York Times*, May 23, 2017, www.nytimes.com/2017/05/23/dining/noma-tulum-pete-wells-mexico-rene-redzepi.html (accessed November 20, 2017).

Part Two

The Identity and Politics of Mexican Foods—and the Politics of Identity

5

A Taste for Agave: The Emerging Practices and Politics of Mezcal Connoisseurship

Ronda L. Brulotte
University of New Mexico

> ***Wahaka Reposado con Gusano*** *Pale silvery straw color. Aromas of cigar, roasted root vegetable, cucumber and delicate spice-herbed rubbed pork with a silky, dryish medium body and a delicately smoky nut, cream, melon, and singed grass finish. Stylish and almost feminine.*
>
> Online review of Wahaka brand *Reposado con Gusano* mezcal[1]

Descriptions like this one of Wahaka's liquor are becoming more common as Mexican mezcal, particularly from the southern state of Oaxaca, gains notoriety in national and international markets. To the uninitiated, wine and spirits reviews may read as over-the-top, almost nonsensical tributes or as mortifying disclaimers. But in the world of prestige beverages such as wine and Scotch whisky, comparisons to seemingly unrelated foods, non-edible flavors and aromas, and the attribution of anthropomorphic characteristics are de rigueur. That Mexican mezcal is now characterized in this fashion indicates that it has entered a particular regime of connoisseurship—one from which it was formerly excluded.

Mezcal, an agave-based distilled alcohol, exemplifies how a once "peasant," indigenous, or working-class food and drink, has now joined the class of celebrated boutique products. Historically, mezcal was, and continues to be, an important element in local community life, particularly as a part of ritual and feast events throughout Oaxaca and other parts of Mexico; celebrations such as baptisms, weddings, and Catholic saints' day celebrations feature mezcal prominently. Until the late 1990s, mezcal was perceived within Mexico as a provincial drink, one produced on a relatively small scale for local consumption. In contrast with tequila, which developed a national and international commercial market beginning in the late nineteenth century, mezcal was virtually unknown outside of Mexico until the 1990s (Gaytán 2014; Orozco 2014). Today, mezcal is still produced on a relatively small scale and is still considered a regional drink by many. However, it has also entered the global export market and is rapidly establishing itself as a quintessential Mexican spirit in culinary taste-making cities such as New York, San Francisco, and Barcelona.

How did a low-status liquor become prestigious among a new class of Mexican and international consumers? Extensive capital investment, certification procedures, regulatory commissions and protected geographical indications, and clever branding campaigns all figure into the equation. However, the recent cultivation of a taste for mezcal also relies on the drink's "interdiscursivity" to create new frameworks of connoisseurship, particularly among North American consumers. Defined as "the use of elements in one discourse and social practice which carry institutional and social meanings from other discourses and social practices" (Candlin and Maley 1997: 212), interdiscursivity encourages us to think through how certain symbolic and material practices associated with the connoisseurship in one realm provide the terms for evaluating taste and value in another. Specifically, how do taste-making practices borrowed from other food and beverages make mezcal legible or relatable outside of its original contexts of production and consumption while simultaneously highlighting its rarified qualities. The examples supporting this transition point to the repositioning of mezcal within a particular race and class politics that underlie Oaxaca's broader development as a "foodie" destination (Johnston and Baumann 2014).

The revaluation of Oaxaca and mezcal

Mezcal functions both as an umbrella term that refers to any beverage distilled from agave and as its own category of agave spirits, apart from tequila, that is regulated by Mexican federal law through its own denomination of origin (DO) (Bowen 2015).[2] Under the first definition, tequila qualifies as a type of mezcal, which, according to law, may be made (1) from only one particular type of agave, *agave tequilana*; and (2) in specifically designated geographic areas, primarily ones located in the state of Jalisco and in limited regions of four other states, Guanajuato, Michoacán, Nayarit, and Tamaulipas. Mezcal under its own DO is produced mainly in the state of Oaxaca and minimally in eight other states (Durango, Guanajuato, Guerrero, San Luis Potosí, Tamaulipas, Zacatecas, Michoacán, and, most recently, Puebla) and may utilize any number of different agave varieties. Thus, while all tequilas are mezcal, not all mezcals may be considered tequila.

This distinction is important because the separate trajectories of tequila and mezcal are very much rooted in Mexico's historically and geographically uneven development (Martínez 2016). In her cultural history of tequila, sociologist Marie Sarita Gaytán (2014) describes how the agave distillate from the Tequila region in Jalisco achieved iconic status as the national drink of Mexico, while Oaxacan mezcal languished as a low-status regional drink. She identifies several factors for this discrepancy. First, the Tequila area's close proximity to and relationship with the economically bustling city of Guadalajara ensured a steady flow of interest and capital investment into Jalisco state. Moreover, Jalisco's extensive, exploitative hacienda system, which historically allowed for the tight control over local labor and land rights, created the stable conditions and infrastructure conducive to a system of modern, efficient alcohol production and distribution. Even then, getting elite Mexicans to judge tequila acceptable to drink required a degree of psychological engineering. Historian José Orozco (2014) chronicles the influential Sauza family in Jalisco, whose patriarch, Cenobio Sauza, was the first maker to export tequila to the United States starting in 1873. Sauza knew that gaining the approval of international, particularly American, consumers would be key to penetrating the middle- and upper-class Mexican market. To this end, Sauza entered his tequila in the 1893 Chicago World's Fair to demonstrate that it was "simultaneously an authentic product of Mexico and a clean, non-disgust-inducing, civilized, and cosmopolitan liquor like cognac, champagne, or brandy" (Orozco 2014: 191). Today, while Sauza may not represent the high-end of the tequila market, it is one of the most recognized liquor brands in the world and is an exemplary "Mexican" produce despite being owned by a multinational liquor company headquartered in Chicago (Martineau 2015).

On the other hand, Mexico's other primary mezcal-producing state, Oaxaca, was geographically isolated and centered on a small, provincial capital whose economic importance had diminished significantly since the colonial period. According to historian William Taylor (1972), Oaxaca's large native population, unlike Jalisco's peasantry, was more economically independent and maintained control over large areas of land. Whereas the hacienda system flourished in much of the country, Oaxacan indigenous communities not only managed most of the region's land and natural resources, but they often also regulated significant sectors of the internal market. Because indigenous residents were perceived as unlikely to embrace the idea of private property, Oaxaca was less attractive to capitalist investors, national and foreign alike. As such, it economically languished throughout the nineteenth and much of the twentieth century. Thus, the reach of Oaxaca's mezcal industry—like its many other rural industries (Cook 2014)— did not extend beyond a regional market to national and international markets the way that Tequila's larger-scale, commercial output did.

Oaxaca's fate as an underdeveloped colonial backwater, ironically, set the stage for it to emerge as Mexico's top travel destination today (Murphy and Stepick 1991). Excluded from the large-scale agricultural and industrial development characteristic of central and northern Mexico, Oaxaca's historic poverty and indigeneity, since the 1970s, have been converted into a folkloric resource employed by its ever-expanding cultural tourism industry. A mix of colonial architecture, colorful markets, and a seemingly endless calendar of religious celebrations attract Mexican nationals and international travelers who are drawn by Oaxaca's promise of well-preserved indigenous culture. Oaxaca is both exotic and "Other," but repackaged in a way that is safe and consumable for outsiders.

Oaxaca's reputation as a repository of authentic culture and tradition has been driven in large part by its craft economy, which has seen the development of traditional craft industries in addition to newer ones such as the brightly painted woodcarvings often referred to as *alebrijes* (see Chibnik 2003). In addition to the wide array of folk arts, local foods and culinary offerings are now centerpieces of the Oaxaca tourism complex and part of a larger "Oaxaca brand" that circulates worldwide (Brulotte 2012; Brulotte and Starkman 2014). Like many things Oaxacan, mezcal is currently in vogue. In his insightful exposé of what he calls the "mezcal rush," Granville Greene implies these categories may coalesce, having heard mezcal described as "drinkable folk art" (2017: 180).

It is also not a coincidence that mezcal's meteoric rise in popularity has coincided with Oaxaca's recognition as a global culinary hotspot where authentic, indigenous tradition is perceived to meld seamlessly with cutting-edge culinary innovation. Mezcal is exemplary of craft food and beverage (see Paxson 2012); with the exception of a handful of larger, industrial brands (e.g., *Zignum*), most Oaxacan mezcal is produced on a relatively small scale, in backyard family operations or distilleries employing a handful of laborers, in which few, if any, production stages are mechanized. Its emerging market is predicated on its image as a traditional, artisanal beverage produced under specific geographic, geological, and climatic conditions, denoted by the French term *terroir* or what anthropologist Amy Trubek has termed "the taste of place" (2009). These factors make mezcal particularly attractive to global consumers' appetites for increasingly specialized food and drink.

The emergent mezcal connoisseurship discussed is directly related to the wider phenomenon of culinary cosmopolitanism, defined as "the desire to seek and experience the cuisines of the Exotic Other" (Johnston and Baumann 2014: 94; see also Heldke 2003). As in other instances where cosmopolitans have a "reputation for being open-minded and elite" when it comes to gastronomy (Cappeliez and Johnston 2013), mezcal connoisseurship is rooted in social taste hierarchies even as it may strive to transcend them.

Acquiring taste and value

As with other forms of alcohol, mezcal might be best described as an "acquired taste." Indeed, one US author who has awarded himself the title of "Mezcal PhD" and maintains a popular website by the same name notes that, "there is a saying in the mezcal world that goes … 'mezcal is an acquired taste, but a taste worth acquiring'" (McEvoy 2013). This presumption refers not only to learning to enjoy mezcal as a sensorial experience but also to understanding the consumption of mezcal as a thoroughly social act. A novice must learn to enjoy the burning sensation of mezcal as it passes through the throat—in part a function of its high alcohol by volume (ABV)[3]—and must come to appreciate the aroma and taste of wood smoke and the flavors of the various types of agave utilized in its manufacture.[4] This individual sensorial experience is always embedded in a wider sociocultural context. Hence, the process by which one actually acquires a taste for agave distillates such as mezcal or tequila may vary.

In the United States tequila often serves as a baseline reference point for those new to mezcal, where many drinkers will have tried a margarita cocktail or downed a shot of tequila at some point. It is not uncommon to hear mezcal likened to "smoky" or "spicy" tequila (see below) as a way of providing context for a first-time mezcal drinker. The experience of the liquor will differ depending on whether the drinker grew up in an Oaxacan community where mezcal is regularly consumed in ritual or other social contexts, or if they first encountered mezcal at a swanky cocktail bar in New York. Certainly, the history of mezcal consumption in local Oaxacan communities is worthy of study in its own right. However, here the focus is on individuals who are socialized in communities where mezcal is *not* traditionally present. In other words, how does one learn to "appreciate" mezcal when the historical or cultural context associated with it is virtually absent?

Mezcal is emblematic of how drinks as embodied semiotic and material forms mediate social life (Manning 2012). Pierre Bourdieu ([1979] 1984) famously theorized that consumer choice reflects a symbolic hierarchy that is determined and preserved by the socially dominant in order to maintain their distinction from other classes of society. Using the consumption preferences of middle-class French society, Bourdieu argued there is no such thing as pure, individual "taste"; all of our different aesthetic choices, from the clothes we wear to the foods and music we consume, are really forms of distinction. Moreover, these "judgments of taste" are an expression of social class formations in modern society—the tastes of one class are formed in opposition to those tastes that characterize other social classes.

For instance, wine is subject to a social organization of value-determining connoisseurship. On one hand, someone who has yet to appreciate a wine's subtleties has also not yet been accepted as part of a recognized social stratum, one in which the drink is regularly consumed and appreciated. On the other hand, if a socially prominent wine drinker learns to appreciate a liquor made primarily by working-class people (i.e., mezcal), then the liquor can be elevated because (1) the socially prominent individual has the language to describe and, thereby, elevate the liquor, and (2) the attraction of the exotic Other increases the likelihood that the socially prominent individual will try and will learn to appreciate the liquor.

Linguistic anthropologist Michael Silverstein (2006) shows how language mediates the relationship between humans and the wine they create as both an agricultural and aesthetic commodity. Through what he terms "oinoglossic registers," or more simply "wine-talk," social relationality is "mediated not directly by wine in its physicochemical or even directly sensorial presence, so much as by the discursive processes of representation of [wine]" (Silverstein 2006: 484). Silverstein likens the speech acts associated with wine connoisseurship to other realms where demonstrating expertise is requisite for appreciation. As with European painting or period furniture, "wine-talk" reveals a structure of authority to judgments that radiate from the representative-declarative acts of some to inform those of others by a kind of interdiscursivity. One quotes or cites or alludes to authority in orienting oneself to the object of aesthetic judgment (Silverstein 2006: 483). In the case of wine, the "proper" authorities may include sommeliers, chefs, or other fellow wine aficionados. Referencing numerical rating systems used by publications such as *Wine Enthusiast*, *Wine Spectator*, and *Wine Advocate* is a way to appeal to higher authority in crafting one's own expertise.

Critically, wine as a drinkable commodity is subject to a host of value-creating judgments at every stage of its "social life" (Appadurai 1986). An agricultural product (the wine grape) is transformed into a prestige consumer good (wine) through a complex value-creation chain (Black and Ulin 2013). In this value regime, wine is recognized as,

> the emergent product of refined, sometimes "boutique" agricultural techniques ... of growing, harvesting, pressing, vat or barrel fermenting, and bottling followed by distribution as at wholesale and retail levels, reaching consumers as a measured, packaged, labeled substance with a history and even a future bound up with numbers: bottle volume in standard sizes, year of vintage (harvesting and final release), quantity of production, number of exemplars still in commodity circulation at any given moment, etc. (Silverstein 2006: 483)

Silverstein's description of this chain reveals that true connoisseurship relies on the mastery of highly specialized knowledge: understanding vintage, grape varietals, growing regions, wine tasting notes, how to pair specific wines to foods, the differences in stemware, etc. While any hobby or profession may have its own specialized knowledge/language, Bourdieu reminds us that not all of these knowledges/languages are created equal. As we will see in the next section, the interdiscursivity—and exclusivity—that Silverstein associates with "wine-speak" transcends wine. It functions within and across different classes of beverages, creating interdiscursive frameworks with other prestige food and drinks, including mezcal.

Value-creation in mezcal

Not all the mezcals consumed as part of the global boom are what might be termed "boutique spirits." Market segmentation exists and is, arguably, necessary to elevating mezcal's social and economic value. Oaxacan mezcal today constitutes a range of products, including noncommercial mezcal that is produced for consumption within households and/or communities of origin as well as relatively inexpensive, commercial brands (e.g., Oro de Oaxaca and Monte Albán) that have been commercially available for decades both within and outside of Mexico.[5] However, the mezcal that *New York Times* food writer and critic Eric Asimov described in 2010 as "tequila's smoky, spicy cousin" and the "Next Big Thing" in the spirits world might be best described as "third-wave," a term borrowed from the coffee industry that seeks to elevate a type of coffee as artisanal rather than as a foodstuff or commodity (Asimov 2010).

Indeed, mezcal's trajectory is not wholly unlike coffee's historical transformation from a mass drink into a specialty product. Anthropologist William Roseberry (1996) described coffee's transformation into a "yuppie" drink in the American market beginning in the 1960s and continuing in to the 1980s and 1990s. Second-wave coffee pioneers Peet's Coffee and Starbucks provided US consumers with an expanded coffee repertoire that included new varieties, roasts, and preparations: Arabica, French roast, espresso, macchiato, etc. All became terms by which one could distinguish oneself from Folgers and Maxwell House drinkers. The more recent third-wave coffee movement, in turn, prizes artisanal coffee, emphasizing small batch or microlot beans that have distinctive tastes tied to particular terroirs (Smith 2018). Additionally, it often takes up various social, economic, and environmental causes associated with coffee production. An ethos of product authenticity and sustainability is critical to such brands, even those that now have a national reach —signaled by terms such as "Fair Trade" and "bird friendly"—making coffee an ideal case study for scholars interested in ethical consumerism (e.g., Cailleba and Casteran 2010).

Beyond the designation for coffee produced by companies such as Counter Culture and Intelligentsia, the term *third-wave* figures as central to value-creation within other classes of specialty products. The "bean-to-bar" chocolate model, for example, relies on the traceability of a raw material (cacao) to a particular national or regional location that supposedly imbues the finished product (chocolate) with a unique flavor profile (Schnepel 2002). Creo Chocolate, a company based in Portland, Oregon, that sources heirloom Arriba Nacional cacao from Ecuador, describes its chocolate in the following manner: "This is a handcrafted bar. We roast, crack, winnow, grind, conche, temper, mold, and package small batches of chocolate in our factory and tasting room. Our cacao beans are sourced directly from the farmer and crafted to bring out their interesting flavors" ("About Creo Chocolate" 2017). Embracing an ethos of environmental sustainability, the company also participates in the Heirloom Cacao Preservation Fund, a US nonprofit organization whose mission is to "save the quickly diminishing *Theobroma cacao* trees that produce the most high [*sic*] quality, flavorful chocolate and identify the farmers who grow them" (see "The Heirloom Cacao Preservation Fund" 2018).

Such specialized language, whether it be "wine-talk," third-wave coffee, or bean-to-bar chocolate, emphasizes commodity as craft and product traceability to particular lands and/or producer over other potential aesthetic or gustatory qualities. Similar elements are at work in the present-day mezcal industry. Thus, a recent article in *The New Yorker* magazine points out that, "The mezcal boom coincides with the popularity of farm-to-table food, the rise of the craft cocktail, and the advent of the bartender as an advocate for environmental and social justice" (Goodyear 2016). *New Yorker* readers are an obvious target market in the United States for third-wave mezcal because the magazine's demographics for its readership include well-educated, prosperous individuals who are extensively traveled and versed in contemporary political and cultural issues. These readers often dwell in major urban areas where mezcal is available or have the means to travel directly to the source in Oaxaca. More importantly, the mezcal boom does not merely coincide with the farm-to-table, craft cocktail, and related artisanal food and beverage movements; the wider framework of "craft" is integral to the appreciation of mezcal: it is to be sipped, savored, debated, and evaluated through the tasting mechanisms like wine and other well-known distillates such as cognac and whisky. Not surprisingly, traceability matters.

In 1994, in an attempt to regulate and legitimize mezcal, Mexico ratified the *Norma Oficial Mexicana* NOM-070-SCFI-1994, the law governing mezcal's denomination of origin (DO). NOM-070 laid out standards for mezcal's production, certification process, labeling, and regional classification system. Modeled on the European system of controlled appellations of origin, a DO in Mexico is defined as "the name of a geographical region of a country that serves to designate a product originating therein, the qualities and characteristics of which are due exclusively to the geographical environment, including natural or human factors" (cited in Bowen 2015: 66). As with tequila, which was accorded first geographic indicator/DO protection in Latin America in 1978, the mezcal DO from the beginning opened up a whole new place-based "rhetoric of authenticity and heritage" (Gaytán 2018) with Oaxaca as its epicenter.

The conception of mezcal as an artisanal spirit was the driving force behind Del Maguey, a company founded in 1995 by Ron Cooper, a New Mexico-based artist. Cooper is generally credited as the first importer of artisanal mezcal to the US market (Greene 2017), where he introduced—and subsequently copyrighted—the idea of Del Maguey's "single village mezcal." The phrase evokes single malt Scotch, whisky produced in Scotland through a pot-still method at a single distillery with malted barley as the only grain ingredient. Unlike the other commercial brands available outside of Oaxaca at the time, Del Maguey's single village mezcals were ostensibly produced in one village, at one *palenque* (mezcal distillery), using one type of agave. Other artisanal brands emerged in the wake of Del Maguey, focusing on small batch production from specific communities or families. Bottles of these mezcals generally start at around $40–50 but may cost upwards of $270 for a bottle distilled with *pechuga de conejo*, or rabbit breast, from Pierde Almas, a company founded by another American artist, Jonathan Barbieri.

While it seems simple enough from a marketing perspective, Del Maguey's model changed the playing field for mezcal in the United States. Consumers now had a conceptual framework for understanding mezcal in much the same way that they had come to appreciate the vast number of wines that were available to them. Cooper, unwittingly or not, had inserted the French concept of terroir into the mezcal world. Today, it is not uncommon to hear the Spanish term "*terruño*" (literally, piece of land or native soil) used to describe mezcal in the same fashion that terroir is used to describe wine. Location—soil conditions, altitude, temperature, water sources, etc.—is critical to third-wave mezcal's popularity. It is not enough to know that mezcal comes from Mexico or even a specific state such as Oaxaca; true mezcal connoisseurship relies on a deeper understanding of Oaxacan geography. Whereas outsider knowledge of Oaxaca was once limited primarily to the state capital, Oaxaca City, its immediate environs, coastal resorts, and previously undetected regions (e.g., Miahuatlán and Sola de Vega) and their smaller communities (e.g., San Luis Amatlán and Candelaria Yegolé) are now on the mezcal aficionados' radar.

Knowledge of terroir is only one aspect of the interdiscursivity that Silverstein identifies with "wine-talk," and the same may be said of mezcal. From plant to product, mezcal authorities elevate the spirit's value through other specialized, yet highly profuse discourses. (Notably, those authorities tend to be distributors, marketers, and consumers or other boosters rather than actual producers.) As with wine, third-wave mezcal holds the promise of unlimited sensorial and intellectual pursuit. In *Mezcal: The History, Craft and Cocktails of the World's Ultimate Artisanal Spirit*, Emma Janzen describes mezcal as "wild and free and (almost totally) unrestricted in how one can approach the production details," adding that its appreciation requires "a more cerebral sense" (2017: 11). Janzen's characterization of mezcal contrasts sharply with its previous image as a harsh, hangover-inducing spirit. She instructs the reader that it should be bottled so "a mix of many flavors comes to a balance" (106), allowing for the "soft, flavor of mezcal" (134) to come through. Drawing on the allure of both cocktail culture and *terruño*, she also waxes poetic about how the same cocktail will taste completely different depending on the region that produced the mezcal.

Nowhere is the enumeration of mezcal's "boutique agricultural techniques" (Silverstein 2006) and production methods more clearly demonstrated than the current generation of bottle labels. The following information taken from a bottle of *Mezcalosfera*, a brand from Mezcaloteca, a by-appointment-only tasting room in the city of Oaxaca, which bills itself as "a project dedicated to the diffusion and conservation of Mexico's traditional mezcal and its diverse biocultural processes" ("MEZCALOTECA" n.d.; my translation)[6] exemplifies this trend:

Tobalá/Madrecuixe

Master Mezcal Maker	Felipe Cortés
State	Oaxaca
Town	Miahuatlán
Maguey utilized	*A. potatorum* 50%, *A. karwinskii* 50%
Maguey altitude	2100–1560 meters above sea level
Oven type	Conical in-ground pit
Wood used	Mesquite and Oak
Grinding method	Stone wheel
Type of fermentation vat	Ashe juniper
Fermentation	Wild yeast
Water used for fermentation	Spring water
Fermentation time	7 days
Number of distillations	2
Type of still	Copper with condenser
Dilution and blending	*Puntas* and spring water
Distillation date	August 2015
Liters produced	220
Lot number	01–15

(my translation)

Not all mezcal labels contain this level of production detail. Nevertheless, providing some combination of these descriptors has become standard labeling practice for artisanal mezcal. Importantly, third-wave mezcals typically include the name of the "master" mezcal maker (*maestro mezcalero*) who made them. Del Maguey founder Ron Cooper not only coined the expression "single village mezcal." He is also strong in his conviction that mezcal's *terruño* is in part attributed to what he calls "the hand of the maker" (Ron Cooper, pers. comm., April 17, 2015). In other words, each bottle is essentially a work of craftmanship that may be traced not only to a single location but also to a single individual. In the same way that the signature of a well-known artist may add value to a piece of Oaxacan folk art, naming the *maestro mezcalero*—or better yet, including his actual signature on the bottle—adds value to the finished product.

While certain production elements are self-explanatory, such as the maker's name and the state or town of the mezcal's origin, other elements like those from the bottle listed above demand higher order connoisseurship for their comprehension. Only the most educated of consumers, for instance, would recognize the scientific names of specific magueys,[7] *A. potatorum* and *A. karwinskii* and their respective local names, *tobalá* and *madrecuixe*. Furthermore, both tobalá and madrecuixe are distinguished from the domesticated *espadín* (*A. angustifolia*), the most commonly used agave in Oaxacan mezcal production. Even though today tobalá and madrecruixe are not exclusively found and harvested in the wild, they are considered part of a class of what mezcal aficionados refer to as "wild" and "rare" agave varieties, some of which are in danger of extinction due to increasing demand.[8] Mezcals made from these "wild" agaves are often perceived as more valuable as well as more complex or nuanced in taste. One mezcal brand owner lamented the trend to a US journalist, stating "A lot of consumers who I talk to have become 'Tobalá drinkers only' or 'Tepeztate drinkers.'" Not surprisingly, his brand, Mezcal El Silencio, focuses primarily on *espadín*.[9]

Part and parcel of the recent phenomenon in which mezcal aficionados profess knowledge of the subtleties of agave varietals is the emergence of mezcal *ensembles*. Unheard of ten, even five, years ago, in the mezcal world, this Spanish term is akin to the English term "blend" as applied to other wine and spirit categories. From the connoisseur's perspective, a successful mezcal *ensemble* strikes harmonious balance between the unique characteristics of each agave variety used in the blend. For example, the following review for a four-agave *ensemble* by Oaxaca-based mezcal brand Koch hints that blending is the redeeming quality of an otherwise ho-hum mezcal.

> Much like with cognac, blending is at the heart of this mezcal. Utilizing four different wild and rare species of agave the blender is looking to heighten the drinking experience by keying up certain strengths and balancing out any shortcomings this mezcal might have. Cirial adds spice and earthiness while Tobalá adds body and sweetness. Lumbre brightens and Tobasiche provides depth and herbal characteristics. ("Koch El Mezcal Ensemble" 2014)

Yet mezcal connoisseurship does not end with being able to discern between agave varieties and knowledge of how to blend them. As the long list of information from the label above suggests, different types of materials utilized at various stages of the production process also contribute to mezcal's endless range of flavors and aromas. Thus, mezcal made in a copper still with a serpentine condenser in general has a different flavor profile than mezcal made in a clay pot still; indeed, the latter is considered a more rustic ("ancestral") technology and now warrants its own subdesignation within the mezcal DO.[10] As with wine or whisky, the materials comprising the vats used for fermentation and/or aging (French or American oak, juniper, etc.) add another flavor variable that may be detected by connoisseurs. If there is still a lingering doubt that the interdiscursivity that characterizes wine appreciation has been transposed to mezcal, one must only consider the bilingual Spanish/English "Mezcal Tasting Wheel" (see Starkman n.d.), which provides colorful, descriptive language for deconstructing the spirit: chemical, woody, and earthy further subdivide into mineral, leafy, nutty, and kippery.[11] These in turn yield another set of comparisons: gun metal, praline, and smoke oysters. Like wine, mezcal is now more than a libation—it is the source of infinite taste possibilities.

Conclusion

The "Mezcal Tasting Wheel" bring us full circle—literally and figuratively—to considering how a drink sometimes described as "gut-rot" by detractors is now seen as a complex drink possessing "aromas of cigar, roasted root vegetable, cucumber and delicate spice-herbed rubbed pork." In many respects, mezcal's rise is analogous to Oaxaca's transformation from a poor, backwater region of Mexico to a trendy destination for art and food lovers. While much of the state remains economically marginalized, cosmopolitan consumers have managed to reframe certain aspects of Oaxaca's traditional craft-based market economy (see Cook and Diskin 1975) within a global movement that elevates the economic and social value of artisanal production; farm-to-table, bean-to-bar, and third-wave, while not typically terms applied to mezcal, create a new framework of value-creation based on certain modes of production and traceability to a land and its people. Within this reframing, the historical underdevelopment or noncommercialization of key Oaxacan industries is precisely what has allowed those same industries to flourish in the new global context. Craft, what once allowed Oaxacan mezcal to languish while tequila rose to fame, is now the word of the moment and a seemingly value-added component of much mezcal.

In this chapter I have shown how a cosmopolitan consumer class creates mezcal's value through interdiscursive means, taking elements from the discourse and social practices associated with wine appreciation and other forms of gastronomic connoisseurship. Ten years ago it would have been unheard of to find a foreigner in Oaxaca searching out a mezcal *Tobasiche* (a wild agave variety) from Ejutla de Crespo in the same way that one might make a pilgrimage to Islay to sample a particular Scotch whisky. Like Silverstein's "wine-talk," the current mezcal-talk has opened up new markets and, arguably, avenues for Oaxaca's cultural appreciation. Indeed, at a recent tasting in Albuquerque, New Mexico, a mezcal promoter commented that US consumers were learning to enjoy mezcal the way they had learned to enjoy wine, but perhaps without the same pretension. He said, laughing, "People always tell me, 'You guys are just like the wine people, except you're not snobs.'" But as we know from the case of wine, chocolate, and other craft foodstuffs, newfound appreciation often comes with a high price tag. The market segmentation that we have seen with wine, chocolate, coffee, and mezcal provides opportunities for new realms of connoisseurship, but at what price?

Notes

1 Review by the Beverage Testing Institute (BTI). BTI is an independent Chicago-based marketing service company for wines, beers and spirits. Founded in 1981 it also sponsors the "World Beer Championship." *Reposado con gusano* refers to a mezcal that is "rested" or aged and includes a "worm" (insect larva) inside the bottle.
2 Known as *Denominación de Origin* (DO), or denomination of origin in English, this system of protection is similar to the French *appellation d'origine contrôlée* (AOC), the certification granted to certain geographical indications for wine and other agricultural products.
3 Most "artisanal" mezcal will have a minimum alcohol content of 45 percent (90 proof), and many register even higher percentages such as 55–56 percent.
4 Unlike tequila, the DO for mezcal does not specify that it must be made of a single type of agave. Indeed, learning to appreciate the flavors of different agave varietals is an important aspect of mezcal connoisseurship.
5 Approximately $20–30 a bottle in the United States, and even less in Mexico. The artisanal mezcals that are gaining traction in the US market, in contrast, are typically $40 and up, with some bottles costing over $100.
6 Mezcaloteca recently began exporting its mezcal to the United States.
7 Maguey is another term for agave frequently used in Oaxaca and other parts of Mexico.
8 Certain agave varieties that were traditionally harvested in the wild are now domesticated or the result of human propagation from wild agaves.
9 Fausto Zapata, cofounder of Mezcal El Silencio, quoted in Signer (2016).
10 The newest changes to the mezcal DO (NOM 70) designate three categories of mezcal, "ancestral," "artisanal," and a third category simply called "mezcal." For

more information on these designations and a discussion of what they mean see Garrone (2017).

11 "Mezcal Tasting Wheel," © Alvin Starkman, Douglas French, and Sergio Inurriagarro.

References

"About Creo Chocolate" (2017), Creo Chocolate, https://creochocolate.com/ (accessed April 8, 2018).

Appadurai, A. (1986), *The Social Life of Things: Commodities in Cultural Perspective*, Cambridge: Cambridge University Press.

Asimov, E. (2010), "Mezcal, Tequila's Smoky, Spicy Cousin," *New York Times*, August 16, 2010. www.nytimes.com/2010/08/18/dining/reviews/18wine.html?pagewanted=all (accessed August 18, 2010).

Black, R. E. and R. C. Ulin, eds. (2013), *Wine and Culture: Vineyard to Glass*, New York: Bloomsbury.

Bourdieu, P. ([1979] 1984), *Distinction: A Social Critique of the Judgment of Taste*, trans. R. Nice, Cambridge, MA: Harvard University Press.

Bowen, S. (2015), *Divided Spirits: Tequila, Mezcal, and the Politics of Production*, Berkeley: University of California Press.

Brulotte, R. L. (2012), *Between Art and Artifact: Archaeological Replicas and Cultural Production in Oaxaca, Mexico*, Austin: University of Texas Press.

Brulotte, R. L. and A. Starkman (2014), "Caldo de Piedra and Claiming Pre-Hispanic Cuisine as Cultural Heritage," in R. L. Brulotte and M. A. Di Giovine, eds., *Edible Identities: Exploring Food as Cultural Heritage*, 109–123, Farnham: Ashgate.

Cailleba, P. and H. Casteran (2010), "Do Ethical Values Work? A Quantitative Study of the Impact of Fair Trade Coffee on Consumer Behavior," *Journal of Business Ethics*, 97 (4): 613–624.

Candlin, C. N. and Y. Maley (1994), "Intertextuality and Interdiscursivity in the Discourse of Alternative Dispute Resolution," in B.-L. Gunnarsson, P. Linell and B. Nordberg, eds., *The Construction of Professional Discourse*, 201–222, London: Longman.

Cappeliez, S. and J. Johnston (2013), "From Meat and Potatoes to Real-deal Rotis: Exploring Everyday Culinary Cosmopolitanism," *Poetics*, 41 (5): 433–455.

Chibnik, M. (2003), *Crafting Tradition: The Making and Marketing of Oaxacan Woodcarvings*, Austin: University of Texas Press.

Cook, S. (2014), *Life, Livelihood, and Civility in Southern Mexico: Oaxaca Valley Communities in History*, Austin: University of Texas Press.

Cook, S. and M. Diskin, eds. (1975), *Markets in Oaxaca*, Austin: University of Texas Press.

Garrone, M. (2017), "NOM 70 is Here—Now We Just Have to Figure Out What that Means," *Mezcalistas*, March 1, 2017, www.mezcalistas.com/nom-70-is-here-now-we-just-have-to-figure-out-what-that-means/ (accessed March 13, 2018).

Gaytán, M. S. (2014), *¡Tequila! Distilling the Spirit of Mexico*, Stanford, CA: Stanford University Press.

Gaytán, M. S. (2018), "The Perils of Protection and the Promise of Authenticity: Tequila, Mezcal, and the Case of NOM 186," *Journal of Rural Studies*, 58: 103–111.

Goodyear, D. (2016), "Mezcal Sunrise: Searching for the Ultimate Artisanal Distillate," *The New Yorker*, April 4, 2016. www.newyorker.com/magazine/2016/04/04/the-mezcal-tour-of-oaxaca (accessed March 13, 2018).

Greene, G. E. (2017), *The Mezcal Rush: Exploration in Agave Country*, Berkeley, CA: Counterpoint.

"The Heirloom Cacao Preservation Fund" (2018), Heirloom Cacao Preservation, https://hcpcacao.org/ (accessed April 8, 2018).

Heldke, L. (2003), *Exotic Appetites: Ruminations of a Food Adventurer*, New York: Routledge.

Janzen, E. (2017), *Mezcal: The History, Craft and Cocktails of the World's Ultimate Artisanal Spirit*, Minneapolis, MN: Voyageur Press.

Johnston, J. and S. Baumann (2014), *Foodies: Democracy and Distinction in the Global Foodscape*, 2nd edn., New York: Routledge.

"Koch El Mezcal Ensemble" (2014), Distiller, https://distiller.com/spirits/koch-el-mezcal-ensemble (accessed March 14, 2018).

Manning, P. (2012), *Semiotics of Drink and Drinking*, New York: Bloomsbury Publishing.

Martineau, C. (2015), *How the Gringos Stole Tequila: The Modern Age of Mexico's Most Traditional Spirit*, Chicago: Chicago Review Press.

Martínez, O. J. (2016), *Mexico's Uneven Development: The Geographical and Historical Context of Inequality*, New York: Routledge.

McEvoy, J. P. (2013), "Ivan Saldana," Mezcal PhD, September 26, 2013, http://mezcalphd.com/tag/ivan-saldana/ (accessed March 14, 2018).

"MEZCALOTECA" (n.d.), Mezcaloteca®, www.mezcaloteca.com (accessed March 15, 2018).

Murphy, A. D. and A. Stepick (1991), *Social Inequality in Oaxaca: A History of Resistance and Change*, Philadelphia: Temple University Press.

Orozco, J. (2014), "Tequila Sauza and the Redemption of Mexico's Vital Fluids, 1873–1970," in G. Pierce and Á. Toxqui, eds., *Alcohol in Latin America: A Social and Cultural History*, 185–209, Tucson: University of Arizona Press.

Paxson, H. (2012), *The Life of Cheese: Crafting Food and Value in America*, Oakland: University of California Press.

Roseberry, W. (1996), "The Rise of Yuppie Coffees and the Reimagination of Class in the United States," *American Anthropologist*, 98 (4): 762–775.

Schnepel, E. M. (2002), "Chocolate: From Bean to Bar," *Gastronomica*, 2 (4): 98–100.

Signer, R. (2016), "How the Mezcal Boom Is Threatening Wild Agave," *Punch*, June 16, 2016, https://punchdrink.com/articles/is-mezcals-popularity-a-threat-to-wild-agave-varietals/ (accessed March 14, 2018).

Silverstein, M. (2006), "Old Wine, New Ethnographic Lexicography," *Annual Review of Anthropology*, 35: 481–496.

Smith, J. (2018), "Coffee Landscapes, Specialty Coffee, Terroir, and Traceability in Costa Rica," *Culture, Agriculture, Food and Environment*, January 9: 1–9.

Starkman, A. (n.d.), "Alvin Starkman - Background and Qualifications," Oaxaca Mezcal Tours, www.oaxacamezcaltours.com/qualifications.html (accessed March 13, 2018).

Taylor, W. B. (1972), *Landlord and Peasant in Colonial Oaxaca*, Stanford, CA: Stanford University Press.

Trubek, A. B. (2009), *A Taste of Place: A Cultural Journey into* Terroir, Berkeley: University of California Press.

6

Making and Changing Yucatecan Taste in Yucatán: Innovation and Persistence in Yucatecan Gastronomy

Steffan Igor Ayora-Diaz
Universidad Autónoma de Yucatán

Introduction

In 2013, three years after UNESCO's inclusion of Mexican cuisine in the list of "Humanity's Intangible Cultural Heritage," the government of Yucatán declared the food of the state as the region's cultural heritage. The government of the state of Yucatán was not alone, as other Mexican states, such as Veracruz and Puebla, had declared their own food as their cultural heritage (Gobierno del Estado de Puebla 2010; Gobierno del Estado de Veracruz 2009; Gobierno del Estado de Yucatán 2013). In the context of these declarations, in Yucatán, the number of restaurants claiming specialization in Yucatecan food has increased. New restaurants have appeared, some owned by local people, some by entrepreneurs migrating into Yucatán from different Mexican regions. The recognition of Mexican cuisine has had at least a three-fold effect on regional Yucatecan cuisine: it has undergone changes to fit the definition of "indigenous," downplaying the global roots of Yucatecan cooking; restaurateurs have introduced changes to renew regional cuisine in accordance to "modern" values and the aesthetics of haute cuisines; and some chefs now affirm the originality and "tradition" of Yucatecan gastronomy as having a cosmopolitan disposition, succeeding at blending local indigenous cooking ingredients and technologies with multiple European and Middle Eastern culinary traditions.

In 2017, I visited a new restaurant twice in the east of the city. The owner, the chief of operations, and the chefs are all of Yucatecan origin. Their restaurant is named after a local, endemic species of birds. On my first visit with relatives and friends, we all ordered different dishes from what appeared to be a "traditional"

Yucatecan menu: we began with black beans and *joroches* (hollow corn dough shaped in small balls around the cook's finger), followed by different main courses: *poc chuc*, a pork filet grilled over wood; *lomitos de Valladolid*, pork loin in tomato sauce; *pavo en escabeche*, turkey stewed in pickling spices; and *mondongo*, stewed tripe. However, when we received our orders, it became evident that the chef had adapted regional recipes to the aesthetics of high cuisine, presenting small servings in imaginative arrangements. Epazote leaves had a strong presence in dishes where this ingredient is regularly more discrete, or usually absent. I was both surprised by the flavors of the tripe I ordered and pleased with the culinary experimentation. However, most of my commensals complained about their meals: "that's not the way it should be served," "I don't recognize the flavors," "these are not the right ingredients."

Two months later we learned from their Facebook page that they had included *mucbil pollos* in their menu on occasion of the days of All Saints and the Day of the Dead. Mucbil pollo is a type of tamale cooked almost exclusively on the last days of October and first days of November (that is, in restaurants—at home, some families cook and eat it only on November 1 and 2, or one week later, for what is locally called *ochavario*, the week after the Day of the Dead). Although we would have usually gone to a different restaurant, established in 1962 in downtown Mérida, a newly invented "tradition" (to fill a street with flower arrangements and sculptures allegoric of the Day of the Dead) prevented us from getting there. Thus, we agreed with friends to meet at this new restaurant to eat this seasonal tamale. To our disappointment they had run out of mucbil pollos. Still, we sat there and ordered a different one (*brazo de reina*, rolled corn *masa* mixed with chopped chaya leaves, and stuffed with ground pumpkin seed and slices of hard boiled eggs), and "Maya" hummus (garbanzo puree topped with roasted squash seeds and sprinkled tomato sauce). The waiter also brought, courtesy of the restaurant, a disk of baked corn dough mixed with achiote, over a bed of thick gravy. When we examined it, we realized that the menu had changed from our previous visit. The meals we ordered triggered contrasting responses: from strong disapproval to praise. Our main course selections at the table were breaded pork filet, turkey in black stuffing, beef in *pipián* sauce, venison tostadas, and baked fish. Like in our previous visit, the aesthetics of the plating favorably impressed everybody but everybody complained about the flavors. While the first time we were there, the opinions were that the flavors were not "right," but were pleasing, this time some disliked the flavors, and one said with a skeptical facial expression and tone of voice: "I don't even understand what is it that I am eating." Nonetheless, one of my friends at the table, a Yucatán-born man, remarked that most Yucatecan restaurateurs seem to lack imagination to innovate "our" gastronomy, and unfortunately outsiders had to be the ones who made new things possible with our "traditional" food.[1]

In this chapter I examine the transformations in the taste of and for Yucatecan food, both within the everyday culinary and the institutionalized gastronomic assemblages, and their affective, political intersections with the transformation of subjective and collective modes of identification with local, regional, ethnic, national, and cosmopolitan cultural values. I argue that the taste of and for Yucatecan food has an important historical, social, and cultural dimension tied to the history of strong regionalism vis-à-vis Mexican nationalism. However, since the end of the twentieth century, the demographic structure of the region has rapidly transformed, and the local-global foodscape has changed, steering the taste for food in multiple directions away from Yucatecan gastronomy. To support this argument, in the first section of this chapter I discuss the sociohistorical and cultural dimensions of the taste of Yucatecan food, and how it has been historically tied to regional identities and politics. In the second section I describe the expansion and fragmentation of the translocal foodscape and the responses of Yucatecans and non-Yucatecans to the multiplicity of food choices, including varying degrees of acceptance and resistance, or rejection to changes. In this context I argue that in contemporary Yucatán we witness, simultaneously, the intensification and the denial of the coextensivity between an established gastronomy and the cultural culinary preferences of the inhabitants of the Yucatecan territory that are leading to meaningful transformations for the taste of and for Yucatecan cuisine.

The taste of Yucatán: The politics of repetition

Research in the fields of physiology and biochemistry suggests that our tongue is an imperfect, limited organ possessing gustative papillae severely restricted to distinguishing among four to six basic flavors (Spence 2017). However, we are aware, from our everyday experience, that the taste of food is much more complex than the sum of those basic biophysiological perceptions (Perullo 2016). In recognition of this complexity, new biological studies are looking at the importance of aromas captured both in the front and in the back of the nose (Shepherd 2011; Stuckey 2012), and the "mouthfeel" perception of the texture of foods (Mouritsen and Styrbæk 2017). Taste, evidently, is a rather more complex experience in which all the senses are implicated at once and, in addition, our social and biographical trajectories play an important part in shaping our total experience of the taste of meals (Howes 2005; Sutton 2001). Furthermore, the meaning of taste in food is supplemented by the existence of aesthetic standards important in distinguishing between different groups of people (ethnic groups, social classes, nations) and in the establishment and reproduction of social hierarchies (Bourdieu 1984; Gronow 1997; Korsmeyer 1999; Lane 2014). Here I take a close look at the mechanism of repetition as important in the process of naturalization of taste. I argue that the taste of and for Yucatecan food is perceived to be the product of a "natural" disposition of the inhabitants of the region toward a certain food aesthetic (i.e., sensual and multisensorial traits of the meals) (see also Ayora-Diaz 2012a, 2012b). Hence the people of Yucatán have, throughout the twentieth century, progressively established a gastronomic code where certain colors, aromas, flavors, and textures are favored over others. Although this is usually perceived as a "natural" inclination, they are the product of a history and a structure of unequal relations between Mexican nationalist and Yucatecan regionalist identities; between indigenous and nonindigenous peoples; between elites and subordinate groups; between "local" people and immigrants; between local and global processes; and between genders.

I have described and discussed in several places (e.g., Ayora-Diaz 2010, 2012a) the historical background of Yucatecan regionalism. In brief: Yucatán was born in 1821 as an autonomous republic, separate from Mexico (formerly New Spain). After joining Mexico as a federation of republics, between 1832 and 1868 Yucatán seceded three times in response to the growing political centralism of central Mexican political elites. At the end of the nineteenth century, the Mexican government occupied the Yucatán peninsula and divided it into two states (Campeche and Yucatán) and a federal territory, Quintana Roo, overseen by the Mexican government (in 1974 Quintana Roo became a state). This partition tamed nationalist and separatist feelings, but a strong sense of regional identity remains to this day.

This political trajectory is important for understanding the development of Yucatecan gastronomy and a specific regional taste.[2] How does Yucatecan food taste? Can we speak of a regional culinary-gastronomic code clearly distinguishable from others? There are ingredients that one finds repeated in different dishes that have become iconic of Yucatecan gastronomy. The sour juice of Seville oranges is used to marinate pork meat before baking *cochinita pibil*, baked piglet, pork steaks, *poc chuc* (grilled pork steaks), ground pork for *longaniza* sausages, turkey for *escabeche* stew, and to squeeze on eggs, longaniza, habanero chili peppers, and poc chuc filets. The odd, bitter, anise-like, strongly aromatic epazote leaves are found in black beans (the preferred beans in the peninsula among people of Yucatecan ancestry), in tomato sauces used to make *pan de cazón* (several layers of tortillas with black beans, dog fish in tomato sauce, and tomato sauce), to make the sauce for *papadzules* (rolled tacos of minced boiled eggs in a sauce of pumpkin seeds and epazote), and, in some cases, to stew pork loin in tomato (*lomitos de Valladolid*). Red onions are used as a garnish in pork and beans (either black or white), poc chuc, cochinita pibil, *panuchos*, fried fish, and in *x'ní pek* ("dog's nose," a salad/sauce made of minced red onions, radish, habanero pepper, cilantro, and the juice of Seville oranges). Allspice, coriander seeds, cloves, cumin, cinnamon, and black pepper are common ingredients in *recados* (spice paste, curry/masala- or mole-like. See below), and white onions and garlic are frequent ingredients in stews. In contrast to Iberian use, instead of pimiento or paprika, Yucatecans use *achiote* (annatto) to impart a red color onto meals, but it also adds a somewhat bitter flavor to the food (central Mexicans use different red peppers to give red color to their sausages).

Bay leaves, cilantro, oregano, thyme, chaya leaves (*cnidoscolus chayamansa*), in addition to epazote, are often used to season and flavor steaks and stews. When meats and tamales are baked, they are wrapped in banana leaves instead of cornhusks or maguey leaves (which is the central Mexican custom), imparting a different aroma and flavor on the meals. Also, the food of Yucatán, other than for a couple of dishes, is not spicy hot. Three hot chili peppers can be found, usually as side dishes at the table for those who like them—*habanero* pepper, *xkat*, and *max* chili pepper. However, especially among the middle and upper classes hot chili peppers are not favored—though neither are they excluded.

In contrast to what Yucatecans and most people over the world understand as Mexican cuisine, in Yucatán people favor pork and fowl over beef; although Yucatecans like and eat different types of cheeses, and consume cream, they are not part of any of the iconic regional recipes (other than stuffed Edam cheese and when fresh cheese is used as a garnish in *codzitos*—fried rolled tortillas with no stuffing, bathed in tomato sauce and garnished with crumbled fresh cheese). Today, tacos are a common way to wrap and eat food in informal contexts, but many still consider them representative of "Mexican" ways of eating.[3] Another usual contrast, cited even by young people, is that Mexican meals (stews and steaks) are made with hot chili peppers, resulting in spicy hot meals, while Yucatecan recipes are generous in their use of spices and condiments, they are not spicy hot and chili peppers are often used only as side dishes.

Following this characterization of Yucatecan food I believe it necessary to consider how this configuration of flavors, aromas, textures, and colors have become the "natural" preference among the people of Yucatán.[4] Different repetitions make this possible. Within some families' everyday life, recipes are reproduced, orally and in embodying practice, among the female members of the family, from one generation to the next. Recipes printed in several cookbooks specialized in Yucatecan cooking repeat cooking prescriptions from one home to another, from one region of the state to another, from one cultural group to another.[5]

Recados (spice pastes) provide another site for repetition and difference. As suggested above, there is one recado that is specific for a recipe. So, turkey in white stuffing, turkey in black stuffing, turkey in *escabeche*, *cochinita pibil*, *puchero* (vegetable and meat stew), and several other recipes, each has a specific recado that calls for a particular configuration of spices. Although less frequent today, there are still families that roast and grind their spices and inscribe their taste preferences in recipes that remain recognizable as the named meal (along with small differences in their taste): i.e., one family's white stuffing may have more or less cinnamon, or their escabeche may have more or less cumin, but they are always found in the meal, repeating the recipe in different homes and regions of the state. However, recent developments such as the industrialization of food, have led to an increasing homogenization in the taste of some dishes, and recipes are repeated with less perceptible differences. Several small industries produce recados on a larger commercial scale, and the flavor of recados becomes the same regardless of the location of the state, or even of other countries in which these recipes are repeated. Differences are thus reduced to which brand each cook or family favors.

There are a number of recipes canonically consecrated both in local and regional cookbooks, as well as in the menus of restaurants. During this process, the culinary-gastronomic code of Yucatecan taste has become inscribed in the body of local people, and their taste preferences reveal the foundations that rule the production and consumption of Yucatecan food. A few examples can illustrate some everyday rules that encode how meals are cooked and eaten: two stews of pork and beans, one with black and the other with white lima beans (*ibes*), are usually cooked with salt, epazote, and white onion. But the citrus juice added on each to get a tart flavor and to enhance the aroma of the meal is different: on black beans one should squeeze lime juice, while on white beans it must be the juice of Seville oranges. If one likes to consume chili peppers, on the former one may add slices of habanero chili pepper, on the latter ground red chili pepper. Other than in some idiosyncratic cases, most Yucatecans would not eat cochinita pibil with black beans, so many Yucatecans are surprised to find *panuchos* (a deep fried tortilla stuffed with black beans) topped with cochinita, while in contrast this has become the usual way to eat panuchos in Mexico City. By comparison, *salbutes* (a deep fried tortilla made from slightly fermented corn dough) can be topped with whatever one wishes to (usually a choice of *picadillo*—ground meat—eggs, lettuce, shrimps, fish, turkey, cochinita pibil, or almost anything). Also, most Yucatecan tamales are consumed without any accompanying sauce, and cream and cheese are not added onto foods (neither are they fried in butter). While at home commensals may expect changes in their meals resulting from a negotiation among family members who like or dislike certain ingredients, and in improvisational response to the seasonal availability or lack thereof of ingredients, at restaurants *most* Yucatecans expect Yucatecan recipes to respect their "original" taste. Clearly, as argued by Sutton (2001), the source of this "original" taste is usually the person's memory of flavors, aromas, colors, and textures associated with biographical experiences of the meal. However, despite these seemingly ingrained preferences, the taste of Yucatecan food has been slowly changing both at restaurants and in domestic kitchens. I discuss these changes in the following section.

The exploding foodscape: Culinary affections and disaffections

The foodscape, both in the city of Mérida and in the state of Yucatán, has dramatically changed since the end of the 1970s (Ayora-Diaz 2012a, 2014). Access to cookbooks, restaurants, ingredients, and different specialized technologies employed to cook recipes from other parts of the country and the world are transforming local reception to foreign tastes, and leading to changes in the taste of local recipes (Ayora-Diaz 2016). The inhabitants of Mérida, like those of cities all over the world, go out to eat. Some may wish to eat Yucatecan recipes they don't know how to cook, or recipes that are highly elaborate and time-consuming. Others go out to try foods of different origins: Chinese, Japanese, Korean, Thai, French, German, Italian, Lebanese, Spanish, Argentinean, Brazilian, Colombian, Cuban, Peruvian; or from different Mexican regions: Campeche, Chiapas, Hidalgo, Jalisco, Nuevo León, Oaxaca, or Tlaxcala. Some establishments serve tacos, tamales, and other fast-food-like meals, but others are well-established restaurants seeking to appeal to middle- and upper-class consumers of regional, ethnic, and international foods. At these places, Yucatecans become familiar with new flavors and different styles of food aesthetics, and some of these are introduced back at home, changing the taste of domestic Yucatecan recipes. Sauces such as soy, English (Worcestershire), and teriyaki have become everyday additions to different domestic meals. For birthday parties or family reunions, a woman may put together celery sticks, carrots, sprouts, and soy sauce, and she can claim to have cooked a "Chinese" meal. At another social gathering, the husband grills sausages and serves them with mustard and beer, and the hosts and guests share a "German" meal (especially if they bought the sausages from the German charcutiere who has established his business for more than a decade, and expanded into two different branches in the wealthier north of the city); and so on. These new ingredients are transferred to other meals and slowly tastes for food begin to change.

In 2000, when I moved back into Yucátan, the main offense to Yucatecan cooking was the foreigners' request to have mucbil pollos stuffed with cheese and ham. In public spaces and local media (central) Mexicans were asked to leave Yucatán's recipes alone, or to return to their place of origin, leaving Yucatecans and their food alone. In 2017, this and other variations of mucbil pollos have become more or less accepted versions, and today, the "deviation" defined as most "outrageous," is the vegan style.[6] In conversation with Yucatecans, they nod with skepticism and sometimes disgust at the mention of mucbil pollos with cheese, but they express utter disbelief at vegan recipes. Gradually, Yucatecan food has become one more type of cooking within an expanding and varied foodscape, and many local consumers of Yucatecan food are now open to experiments in restaurant recipes—while in the past, experimentation was encouraged at home but rejected in restaurants.

In addition to restaurants, some further transformations have been important in steering these changes. As the number of restaurants claiming Yucatecan food as their specialization grows, I have found that chefs leading the kitchen are often Yucatecans who have attended culinary schools in central Mexico. Upon their return they choose to change some recipes, using a spice or some other element proper to the cooking of central Mexico as their "signature" ingredient. Other chefs trained in central Mexico have adopted the values of fusion cuisine and seek to blend sometimes complementary, sometimes incompatible cuisines. For example, at one restaurant, for several years, the chef created successful fusions of Mediterranean and Yucatecan ingredients (though he never claimed to be doing Yucatecan food); at another, the chef blends the flavors and aromas of Oaxaca and Yucatán; while in another a young chef failed in her experiment of lasagna of cochinita pibil. In addition to these chefs trained outside Yucatán, I also find that now cooks are locally trained in a growing number of culinary schools hosted by private universities. These cooks/chefs come out from school and seek to innovate and experiment to carve their own niche. Lastly, entrepreneurs from other parts of Mexico have decided to change Yucatecan recipes to tailor them to the food preferences of tourists from other parts of Mexico. They have decided to assimilate, for example, papadzules to the logic of Mexican enchiladas and prepare the pumpkin seed sauce with cream; add cheese in recipes that formerly did not have any; or add chili peppers that are alien to Yucatecan cuisine.

The global market of edible commodities has also turned some ingredients, formerly foreign to the region, into everyday ones. Forty years ago, chipotle, jalapeño, and Serrano peppers were available in canned presentations. Today they are found fresh or dried at almost any supermarket. Street vendors roam the streets selling "Mexican" tamales and *tepache* (fermented fruits, especially pineapple, common in other Mexican regions). The pads and fruits of prickly pear are also available at supermarkets, though until recently they were not part of the regional diet. Some chili peppers such as guajillo and ancho were introduced to flavor seafood and sauces, and now are added into Yucatecan meals. Combinations of ingredients that were not part of the Yucatecan code are now turned into everyday preferences. For example, *panuchos* topped with cochinita pibil or with turkey in black stuffing are regularly listed in restaurant menus. Small eateries (*loncherías*) that specialized in the sale of local panuchos, salbutes, and tamales, now additionally sell *sopes* and *flautas*, borrowed from the central and northern Mexican culinary repertoires. As people develop a wider taste for foods, their affection toward Yucatecan cuisine is gradually diminished. Regarding food preferences, many prefer to be seen as "cosmopolitan" rather than as "Yucatecans." As they accept changes in the taste of Yucatecan food, their taste for it changes too. With these, I have also found, there are changes in the relationship between Yucatecan food and identity.

However, not all accept these changes. After a meal, one friend affirmed: "When I go out to eat Yucatecan food, I want them by the letter. I don't want fusion, I don't want experiments!" Like her, other friends are skeptical of other experimental approaches to Yucatecan recipes. One restaurant, in which the owner has placed a food lab—which a chef told me is worth about 30 million pesos or US$150,000, at the current 2018 exchange rate—has been transforming sauces and meals according to the principles of molecular cuisine. Among those who have gone there to try the food I have found some who said it was an assortment of "interesting" renditions of Yucatecan cuisine, while others have told me that the food was not at all worth the price they paid (about US$50 per person, fixed price tag), or plainly, that it was not good at all. Another friend grumbled about a downtown Mérida restaurant run by a Yucatecan chef, in which she said that when she expressed dissatisfaction about one meal, she was told that it was "fusion."[7]

Other Yucatecans, on the other hand, find some experiments intriguing. While in the opinion of many local people, the creation of lasagna stuffed with cochinita pibil is a highly extravagant experiment, some say that even if they have not tried it, it "sounds appealing"; or one friend who tried it, said that she is *so* open to experiments that she liked it. In general, however, I have found that the issue for many local people is not whether the food would necessarily taste bad but is instead that Yucatecan taste should be respected and preserved over experimentation. As my friends say, after all, Yucatecan food has been seen, for several decades, as a "cosmopolitan" food resulting from the ingenuity and creativity of local cooks who were able to adapt Caribbean, European, and Middle Eastern recipes to locally available ingredients. Hence, Yucatecan cuisine and gastronomy were born from gradually introducing foreign recipes into regional cuisine, making it (in contrast to the Mexican discourse that favors the importance of indigenous culinary traditions) a cooking style marked by a food aesthetic that joyfully combines diverse cultural culinary traditions, and claiming to be "Yucatecan" rather than indigenous, Mexican, or Caribbean (despite their obvious historical connections to these different traditions). In fact, the CEO of the restaurant with which I began this chapter, says that their restaurant is "*mestizo*" rather than "Yucatecan," and seeks to blend regional food with imported, new ingredients, seeking new interpretations and renditions of regional food.

These responses suggest that, at the same time that local practices continue to reinforce the coextensiveness of a set of culinary practices that sustain Yucatecan taste, other processes are challenging this coextension, supporting the production of different modes of identification with the Mexican nation, or the world at large. From the viewpoint of the former, the latter are subject to a form of what Homi Bhabha (1994) called "colonial mimesis"; that is, the attempt of local subjects to become more like the subjects of the Mexican colonialist center. From the viewpoint of the latter, the former are carriers of archaic forms of social and cultural parochialism. Here, the notion of *gastronomadism* becomes conceptually important for analyzing culinary changes.

Gastronomadism and the politics of mobile identifications

Today, I find both at restaurants and homes different but supplementary tendencies to change the way in which Yucatecan recipes are followed. Some, like most described so far in this chapter, belong to the sphere of "forward"-looking innovation: cooks adding new ingredients, experimenting with different forms of fusion, or introducing new techniques into the preparation of meals (sous-vide, torches, liquid nitrogen, and making mousses and foams of everything). However, there are other changes that could be called *retro-innovation*; that is, the recuperation of abandoned techniques and technologies such as *molcajetes*, displacing the more "modern" blenders, the readoption of "natural" and "organic" ingredients seeking to replace processed foods on account of some ideal reconstruction of the past, and the imagination of meals before the arrival of Europeans as a model for "authentic" Mayan cooking, to mention just a few of these choices. The seemingly paradoxical term of retro-innovation is here adopted to indicate that these practices cannot lead to the flavors of the past—unless it is an imaginary one—but instead it steers toward new cooking alternatives that enrich the contemporary foodscape.

In this context it is important to conceptualize different forms of gastro-sedentarism and gastro-nomadism. I have argued elsewhere (Ayora-Diaz 2017) that this is a reflexive style used to approach the cuisines one finds available in the contemporary foodscape. Contemporary urban spaces encompass a fluctuating but growing multiplicity of cuisines that tend to relativize the value of any single gastronomic tradition. While in the recent past it was possible to use food preferences to assert individual identification with gender, ethnic, local, regional, or national groups, today food has lost or, in some contexts, is losing this value. Hence, when people affirm their passion for local foods they do so in a social environment in which they are comparing their preferred cuisine to others available at the time. Thus, this is a reflexive choice. In contrast, other people have shifting preferences, both as cooks and as consumers. This gastro-nomadic disposition affirms a cosmopolitan set of values that reposition the consumer against those who choose local meals.

The power of food to mobilize affects among consumers depends on the self-definition of individuals who may choose to stress their love for what is considered local, regional food, or their self-definition as cosmopolitan, rejecting the importance of food to mark one's ethnic, regional, or national identification. These are not fixed and stable sets of preferences. As anthropologists have shown throughout the history of the discipline, identifications are shifting and fluid, depending on contextual political and other social changes. As I have demonstrated in this chapter, contemporary Yucatán is a social space in which cultural practices are assembled into different forms of culinary preference, changes in cooking techniques and recipes, and the introduction of different forms of innovation. I have also sought to highlight the importance of looking at different forms of gastro-sedentarism and gastro-nomadism in contemporary societies, as they are an indicator of the changing cultural politics of identity in the global market of ethnic and national foods.

Notes

1 As already noted, they are all Yucatecan. Yet, the chef who originally designed the menu had worked at high-end hotel restaurants in the United States. Their goal, I was told, was to innovate without radically changing Yucatecan recipes and thus bring young Yucatecans back to the food of their parents. Still, my friend's assumption is telling of a common view, even among local people, that Yucatecans are conservative.

2 As I have discussed elsewhere in great detail (Ayora-Diaz 2010, 2012a, 2012b), during the nineteenth century the Mexican navy blocked the entry of ships from Yucatán into Mexican ports, and of Mexican ships into Yucatán, and Yucatecan elites turned to Europe, the United States, and the Caribbean in search of commercial and cultural ties. The success of henequen plantations strengthened the regional economy and allowed a vibrant commerce of cooking ingredients and technologies from overseas giving birth to new regional and national cuisines.

3 Often, in the Anthropology of Food and Anthropology of the Senses and other courses I teach at the Autonomous University of Yucatán, I have asked students which foods do they know are Mexican. Very often, the first item listed is beef "tacos," and some expand the list adding *sopes* (a tortilla topped with beans and cheese) and *pozole* (hominy stew with beef).

4 "The people of Yucatán" is a problematic expression. The Mayan people inhabited the peninsula long before the arrival of the Spaniards. But people of Iberian ancestry have lived in Yucatán for over 500 years. For more than a century Yucatán has received groups of people coming from different cultures: Chinese, Korean, Cuban (who are, in turn, another ethnically mixed population), French, Germans, Italians, members of the former Ottoman Empire (today's Syrians, Lebanese, Palestinians), and in recent years growing numbers of "expats" from the United States, Canada, different European countries, and from different Mexican regions, some more visible than others. Consequently, here I am using "people of Yucatán"

to refer to those who acknowledge themselves as Yucatecans. Many people from other Mexican regions claim to be Yucatecan after some twenty years of residence, but they keep Mexico City as their main frame of reference for contrasting, usually in favorable ways, against Yucatecan cultural practices (see Ayora-Diaz 2012a; Vargas-Cetina 2017).

5 See my discussion of cookbooks in Ayora-Diaz (2012a). Bak-Geller Corona (2013) has also examined the importance of cookbooks for Mexican nationalist ideology. See also this volume.

6 A newspaper article from 2016 mentions four different types of this tamale as well established and accepted. Among these, two are firmly established variations, and the list adds an (acceptable) third one stuffed with vegetables, while the writer qualifies as "extravagant" the fourth one, filled with ham and cheese. The writer adds: "remember, these are just some of the recipes, a mucbil pollo can [be stuffed] with whatever you wish" ("Cuatro diferentes formas de preparer el tradicional pib" 2016). This is a statement with which many Yucatecans would still disagree today.

7 This was, of course, a disrespectful response, as the customer felt she was put down as "unsophisticated" and unable to appreciate the chef's experimental disposition.

References

Ayora-Diaz, S. I. (2010), "Regionalism and the Institution of the Yucatecan Gastronomic Field," *Food, Culture, and Society*, 13 (3): 397–420.

Ayora-Diaz, S. I. (2012a), *Foodscapes, Foodfields and Identities in Yucatán*, Amsterdam: CEDLA, New York: Berghahn.

Ayora Diaz, S. I. (2012b), "Gastronomic Inventions and the Aesthetics of Regional Food: The Naturalization of Yucatecan Taste," *Etnofoor*, 24 (2): 57–76.

Ayora-Diaz, S. I. (2014), "Lo postnacional y la fragmentación del paisaje culinario yucateco. Transformaciones contemporáneas en los hábitos culinarios," in A. López-Espinosa and C. R. Magaña González, eds., *Hábitos Alimentarios. Psicobiología y socioantropología de la alimentación*, 49–62, Mexico City: McGraw Hill.

Ayora-Diaz, S. I. (2016), "Home Kitchens: Techniques, Technologies, and the Transformation of Culinary Affectivity in Yucatán," in S. I. Ayora-Diaz, ed., *Cooking Technology: Transformations in Culinary Practice in Mexico and Latin America*, 85–98, London: Bloomsbury Academic.

Ayora-Diaz, S. I. (2017), "Gastronomadismo y cultura culinaria. Transformaciones tecnológicas, representaciones y performances afectivos de la identidad yucateca," in A. Guzmán, R. Díaz Cruz and A. W. Johnson, eds., *Dilemas de la representación. Presencias, performance, poder*, 255–283, Mexico City: UAM, INAH, Juan Pablo Editores.

Bak-Geller Corona, S. (2013), "Narrativas deleitosas de la nación. Los primeros libros de cocina en México (1830–1890)," *Desacatos*, 43: 31–44.

Bhabha, H. (1994), *The Location of Culture*, London: Routledge.

Bourdieu, P. (1984), *Distinction: A Social Critique of the Judgment of Taste*, Cambridge, MA: Harvard University Press.

"Cuatro diferentes formas de preparer el tradicional pib" (2016), *Diario de Yucatán*, October 27, 2016, Section "Mérida," online newspaper version. http://yucatan.com.mx/merida/ciudadanos/cuatro-diferentes-formas-de-preparar-el-tradicional-pib (accessed October 13, 2017).

Gobierno del Estado de Puebla (2010), "Decreto del Ejecutivo del Estado, por el que Declara Patrimonio Cultural Intangible del Estado de Puebla a la Gastronomía Poblana," Puebla: Gobierno del Estado de Puebla, March 3.

Gobierno del Estado de Veracruz (2009), "Decreto que Declara la Gastronomía Veracruzana como Patrimonio Cultural del Estado de Veracruz," Veracruz: Gobierno del Estado de Veracruz, July 21.

Gobierno del Estado de Yucatán (2013), "Decreto por el que se Declara la Gastronomía Yucateca 'Patrimonio Cultural Intangible del Estado de Yucatán', " *Diario Oficial*, Year CXVI, No. 32,381, July 14.

Gronow, J. (1997), *The Sociology of Taste*, London: Routledge.

Howes, D. (2005), "HYPERSTHESIA, or the Sensual Logic of Late Capitalism," in D. Howes, ed., *Empire of the Senses: The Sensual Culture Reader*, 281–303, Oxford: Berg.

Korsmeyer, C. (1999), *Making Sense of Taste*, Ithaca, NY: Cornell University Press.

Lane, C. (2014), *The Cultivation of Taste: Chefs and the Organization of Fine Dining*, Oxford: Oxford University Press.

Mouritsen, O. and K. Styrbæk (2017), *Mouthfeel: How Texture Makes Taste*, New York: Columbia University Press.

Perullo, N. (2016), *Taste as Experience: The Philosophy and Aesthetics of Food*, New York: Columbia University Press.

Shepherd, G. M. (2011), *Neurogastronomy: How the Brain Creates Flavor and Why It Matters*, New York: Columbia University Press, Kindle Edition.

Spence, C. (2017), *Gastrophysics: The New Science of Eating*, New York: Viking.

Stuckey, B. (2012), *Taste: What You're Missing: The Passionate Eater's Guide to Why Good Food Tastes Good*, New York: Free Press.

Sutton, D. E. (2001), *In Remembrance of Repasts: An Anthropology of Food and Memory*, Oxford: Berg.

Vargas-Cetina, G. (2017), *Beautiful Politics of Music: Trova in Yucatán, Mexico*, Tuscaloosa: University of Alabama Press.

7

The Life Delicious: Taste and Politics in Mérida, Yucatán

Gabriela Vargas-Cetina
Universidad Autónoma de Yucatán

Food, sound, and sociability

In 2017, my husband and I visited a friend who had returned from a one-week vacation in Cuba. He told us that he and his wife had arrived in Havana on a Friday evening. The hotel offered them drinks and snacks, but he and his wife preferred to go straight to dinner. He proceeded to describe the meal they had been served: the rice was too dry and the lobster over-cooked. Then he described every single meal he and his wife had eaten during the entire trip, telling us which ones they liked and which ones they did not, and why, comparing them to similar meals he has had in Yucatán or in his travels in the United States. We listened to the careful recounting of each breakfast, lunch, and dinner. He then told us there had been "a lot of music all the time" and described it along rather generic lines: cabaret music, guitar trio music, music with electrical instruments, and "tropical music" (which encompasses most Latin American music with conga drums). He never told us about the sights or, beyond the names of the places he visited, anything about the locations outside the restaurants, bars, and food stands he and his wife visited. He did tell us, however, that he had enjoyed the company of the other Yucatecans who had travelled to Cuba on the same trip, organized by a local tour agency, so in spite of some bad meals the overall experience had been good.

Caroline Korsmeyer (1999) chronicles how in philosophy the sense of taste has been put at the bottom of a hierarchy of the senses, a position it shares with smell and touch. Sight and hearing, instead, have been considered the highest senses and the ones that actually engage the intellect, while the other three engage the body as a whole but not necessarily the thinking faculties (Korsmeyer 1999: 11–37).

I am sure most Yucatecans, or at least the ones I know, would take exception to this! Food in Yucatán seems to be everyone's obsession (including mine) across all social classes. Second to food come drinks, and after these come sound and music. And then come the sights. A friend who came from Chiapas some ten years ago was astounded that one of the best restaurants in the city featured live Yucatecan music every day but had the view of the restaurant's parking lot. She said she loved the food and the music but observed that the restaurant should have put up ornaments and plants to obstruct the view of the cars. Today, this restaurant, mainly patronized by Yucatecans, has not changed its décor; it continues to be one of the best in the city; to have a live Yucatecan trova trio as its musical hallmark, to be crowded on Mondays and during most Yucatecan holidays; and to have the view of its parking lot through large windows. After all, for Méridans, the food, music, and ambiance are more important than the sights, and things only make sense in the context of others' company.

There is in Mérida a yearly cycle of parties and celebration that creates the illusion that everything is always well and wonderful. This is because being part of events and celebrations is a way of dealing not only with the happy occasions but also with personal anguish, loss, and even mourning. The general feeling is that one should be with others as frequently as possible in order to have a good life. Besides, the consistently low statistics of crime in the city and the state continue to sustain the mirage of Mérida as a place where life is peaceful, enjoyable, and grief-free. However, no paradise is free of trouble, and right now the threats looming large over Mérida and its carefree-like style of living and partying seem to come both from without and from within. In terms of Jacques Attali's theory of noise (1977), the music is beginning to be disrupted.

Living the life delicious

When I told a group of close friends and relatives that I would be writing a paper called "The Life Delicious"[1] and that it would be about our life in Yucatán, they all said "What a perfect title! That is us and our life! Everything always has to be delicious! The food, the music, the company, the conversation!" Gatherings including food and music are a constant feature of life for Méridans, who take pride in the busy social calendar. The intense sociability extends to the celebrations around individuals and the family life cycle, and to the activities of hundreds of social clubs, sports circles, religious and professional associations, bringing individuals together almost every month.

According to the Mexican Observatory of Policy for Social Development (Consejo Nacional para la Evaluación Pública del Desarrollo Social or CONEVAL) the state of Yucatán continues to be one of the poorest in Mexico, since 42 percent of its population fell within the classification of either "poor" or "extremely poor" in 2016. However, also according to CONEVAL, the municipality of Mérida is an oasis of relative wealth, where over 93 percent of the local population was over the poverty line in 2016 (CONEVAL 2018). Part of the redistribution of wealth in the city no doubt comes from the social activities in which Méridans so happily engage year-round.

The space of the city of Mérida and its coastal offshoot Progreso, along with the grounds of the annual fair in the village of Xmatkuil, are the main areas of the playground within which the social life of Méridans takes place (see Map 7.1). Restaurants, bars, private houses, and spaces called *Salas de Fiestas* (fiesta halls) are an essential part of this sociality. These latter spaces are built or remade with parties in mind; as the sizes of homes and kitchens shrink in

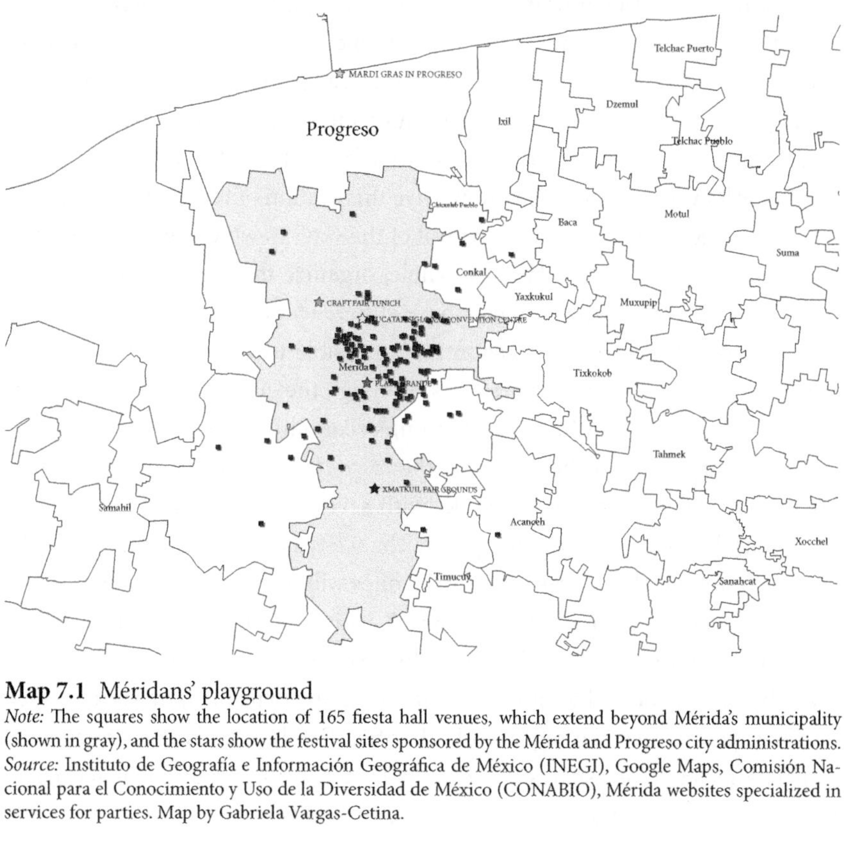

Map 7.1 Méridans' playground

Note: The squares show the location of 165 fiesta hall venues, which extend beyond Mérida's municipality (shown in gray), and the stars show the festival sites sponsored by the Mérida and Progreso city administrations. *Source:* Instituto de Geografía e Información Geográfica de México (INEGI), Google Maps, Comisión Nacional para el Conocimiento y Uso de la Diversidad de México (CONABIO), Mérida websites specialized in services for parties. Map by Gabriela Vargas-Cetina.

Mérida, these venues grow in numbers and in infrastructure. An entire fiesta service sector has flourished around these venues, including catering services, live music, personal beauty specialists, party toys, and photo services. Music is a necessary part of the life delicious among middle- and upper-class Yucatecans, and so is constant talk and, sometimes, shouting. It would be unthinkable for young Méridans to have a night at the disco without shouting out loud the lyrics of the songs, even the ones they don't know or don't understand. Young and old Méridans across all social classes partake in the intense social life of the city in different manners and spaces, but the economy, social demands, and politics of partying year-round touches practically everyone's life across the city. And food is at the center of "The Life Delicious" almost every time.

In February 2018, a group of my and my husband's friends, including two academics from Mexico City, had gathered at a restaurant in Plaza de Santa Lucia in downtown Mérida. One of our friends had presented my book *Beautiful Politics of Music* (Vargas-Cetina 2017) at the International Reading Fair of Yucatán (FILEY) and was explaining what it was about to my sisters and other friends. She said that I describe how in Mérida people are part of myriad organizations. One of my childhood friends, a local entrepreneur who is a prominent member of the Yucatán Chapter of the Mexican Women Entrepreneurs Civic Association, said: "You see, we have all these organizations not precisely because we love to organize; we organize because [we] love to party … Here the party never ends." I will now let my friend's explanation drive this section of this chapter, since her account is a good and accurate portrayal of the ways in which people in the city, and to some extent in Yucatán as a whole, organize their social calendar. Her account glosses over social and economic differences among social groups but, taking these differences into account, it is possible to affirm that these events repeat across the different social spaces, although the quality and prices of foods and drinks change according to each group's economic means.

"In January, we start the year with the *rosca de reyes* [Three Kings' Bread]," my friend said. This is a circular cake with several little toy dolls, called *niños* (children), hidden within the dough. By chance two or three people will find a *niño* each in their slice of cake. The lucky ones will then have to buy tamales for those in the group, in or around February 2, which is Candlemas in the Catholic calendar. Of course, there will be at least two parties resulting from this one, each given by one of the lucky finders of a *niño*. As she explained, in Yucatán each person is part of several groups: work, charities, family, friends, and professional associations. First, one shares the *rosca* with each group, around January 6, and later one shares tamales with those groups, around February 2.

Since at least the mid-1800s, Mérida has been known for its Mardi Gras festivities during the winter, which ever since then have involved most every sector of the local population (Martín Briceño 2014; Vargas-Cetina 2017). At the end of the 1800s and beginning of the 1900s there used to be a competition among social clubs to organize and host the most lavish balls, feature the best music, and showcase the most spectacular *carros alegóricos* (floats) during the three carnival parades on consecutive days. This tradition continued into the twenty-first century, but now the public floats competitions and parades are organized by working-class clubs and groups. Until 2013, the parades took place on the Paseo de Montejo Avenue, in northern downtown Mérida. In 2014, however, the Mardi Gras celebrations were moved to the grounds of the annual cattle and agriculture fair, in the southern city district of Xmatkuil. The social stratification during Mardi Gras became more evident: today most middle-class and well-to-do families abstain from attending the city-sponsored popular festivities, preferring to celebrate their own pre-Mardi Gras balls or, if their children insist, going to the port of Progreso, on the northern coast, where a crowd of all social classes mix. In contrast, *Mardi Gras Plaza* (Plaza Carnaval) is a popular entertainment option mostly for working classes. Mérida's municipality offers free transport, and hundreds of people who have rehearsed their dances and created their own and their family's costumes for the occasion attend the parades, eat at the food stands, and enjoy themselves in the company of their family and friends. Although in segregated spaces, then, Mardi Gras effectively continues to involve most Yucatecans, so the party does go on.

Forty days after Mardi Gras, the elementary schools of Yucatán give children two full weeks of vacations, coinciding with Holy Week and Easter. Later on, during July and August, they give the children two full months off. This latter period is called by Yucatecans *la temporada* (the season). Each time the school board attempts to shorten the school vacation periods there are protests and complaint movements organized by the owners of businesses along the Yucatecan coast, since they derive most of their income from the seasonal escape of Yucatecans during the holidays. Up until the 1980s, it was expected for many middle- and upper-class Méridan families to relocate full-time to the Yucatecan coast for two weeks at the beginning of spring and then two months during the summer. Sisal, Chelem, Progreso, Chibxulub, San Benito, and Telchac are the main coastal settlements where Méridans own summer homes.[2] Progreso is, by far, the main spot for socializing. Its infrastructure of restaurants, bars, hotels, and urban services is second to none along the Yucatán coast. Yucatecans from other parts of the state also visit the beaches, but while day-visitors flock to the *malecón* boardwalk in Progreso and other popular beaches, wealthy and middle-class Yucatecans, who are mostly Méridans, stay in private beach residences and visit and socialize with their friends. Still today, many Mérida restaurants and business establishments move full-time to Progreso or surrounding coastal towns during much of Holy Week, Easter Week, July, and August.

After Easter, usually in March or April, everyone is home, so the celebrations move back to the city. On April 30 it is Children's Day, on May 1 it is Labor Day, and on May 10 it is Mother's Day. And on May 15, it is Teachers' Day. Children are off school on these festive days and the family either stays at home or, if there is a long weekend, goes back to the beach, or organizes large get-togethers with friends. Also, during May, parishes organize daily ceremonies called the "Presentation of Flowers." Children, dressed in white, go to church with May flowers (*plumeria alba*) in baskets to create flower arrangements at the altar, under the proud gaze of their parents and relatives. After the ceremony many families get together at the parish to move the party somewhere else.

Since 1928, Mérida's League of Civic Action made May 10 the official day to honor motherhood—probably, as it has been suggested, to counteract local feminist ideas that saw motherhood as a burden rather than as an important role for women (Vargas-Cetina 2017: 145). Every May 10, families get together to celebrate the mothers in their midst. Those who can afford it go to restaurants, others go to a small and trusted eatery, and still others celebrate at home. In addition, Méridan local groups are well disposed to give awards to their members for different reasons. Private schools, Rotary Clubs, The Lion's Club chapters, gourmets' societies, social clubs of different types, and literary societies constantly organize banquets and award ceremonies, and many of them grant Excellence in Teaching awards to chosen teachers, on or around May 15, Teachers' Day in Mexico. Also, weddings in May take advantage of the return of Méridans to the city. In June the graduation parties start and, as my friend explained, one goes to as many as one's own children, organized by their friends or their friends' parents. After this, it is vacation time again. Many people, as my friend said, "shut down their houses" and move to the beach for two months, where they enjoy the beaches and local restaurants, as well as their beach homes and those of their relatives and friends.

Some middle-class and upper-class families keep small fishing boats or yachts that they use especially during the *temporada*. In the afternoon they put on their swimsuits and get on their boats, to sail to the houses of other friends or to prearranged points where they all meet. They park their boats, then float on the water or stand close to the shore while they talk, drink and eat snacks, and then get back home on their boats to their respective beach houses. People from other parts of Mexico and abroad often see this as a waste of expensive boat fuel, but for Yucatecans the main attraction of having a boat is not just to sail and enjoy marine life but to visit other people, even if they are those you see regularly the rest of the year or have recently visited by car.

At the end of August people return to the city, so that come September everything resumes as usual. Parties also get underway around September 16, to celebrate Mexican independence. In the words of my friend:

> In September, we all become Mexicans. Never mind that the rest of the year we live in our own independent Republic of Yucatán. In September you have to eat pozole, chiles rellenos, [you] dress up as *Chinas Poblanas* and *Charros*, you wave the Mexican flag. You celebrate Mexico with your colleagues at work, with your social club and your groups of friends, in one Mexican night after the other. You listen to Mariachi music. After that, in October, you can go back to being Yucatecan.

During most of October Méridans have pre-*pibes* and then the *pibes* parties (*pibes* are the ceremonial tamales made for the Day of the Dead festivities. See Ayora-Diaz 2012: 136–137). This is because, as my friend explained, one has to practice making *pibes* so they can come out perfectly during the Day of the Dead. In November, Méridans, this time including members of the middle and upper classes, flock to the adjacent city district of Xmatkuil, where the regional cattle and agricultural fair has now become a major yearly event involving rides, concerts, entire streets of stands, and many other attractions. Again, the city of Mérida makes free bus rides available to all Méridans between the city and the fair's grounds, so even those who don't have private cars can attend the festivals. In December people organize *pre-posadas* and *posadas*. These are parties originally inspired by the religious theme of Mary and Joseph asking for *posada* (shelter for the night), but have now fully developed into musical and eating fests organized by groups of friends, neighbor associations, parishes, social clubs, literary societies, and schools. Each group organizes its own posada, either at a fiesta hall or at a local restaurant-bar. The point is to eat, drink, and celebrate, preferably with music. In these instances, working-class people may eat tamales and rice pudding, and drink sodas and rum. In contrast, the middle and upper classes hire cooks to serve *tacos al pastor*, or grilled meats, in addition to or instead of the tamales, and drink whisky, rum, or brandy with soda.

The year closes with the Christmas season when families and extended families organize parties, host friends, and then go to visit the houses of other families. Then come the New Year's Eve balls. These are organized by social clubs and local restaurants, because Yucatecans consider Christmas a family celebration and New Year's a social occasion to eat, dance, and party with one's friends. Of course, besides all these yearly celebrations, there are those related to the more individualized cycle of birthdays, gatherings of extended families, *Quinceañera* balls, and related "Presentation in Society" balls in different social clubs, and other special occasions requiring special parties. These celebrations are turned into spaces and moments in which families display their taste for dresses, foods, drinks, and forms of conviviality. Consequently, families with larger incomes display expensive objects and those who come from more urban families observe etiquette rules that contrast with those of less well-to-do families, especially if they are of rural origins.

Additionally, there are dozens of established festivals organized by local parishes (the Fiesta de San Sebastian being one of the major ones in town) and by the state and city authorities. In downtown Mérida, every week citizens and tourists alike can attend "Jarana Monday's" of Yucatecan Jarana dance performances, "Trova Tuesday" music festivals, the "Thursday Serenade," "Cultural Fridays" performances at the Cultural Center of the Autonomous University of Yucatán, and the free Sunday concert of the Yucatán Symphony Orchestra at noon. A weekend fair in downtown Mérida starts every Friday night and finishes on Sunday night, and it includes "Mexican Night" on Saturday, and on Sunday the Mérida weekly family fair and flea market. Again, different social classes mingle by coinciding in downtown Mérida but across a range of segregated spaces and events.

In March, in addition to the FILEY, which offers free book presentations, symposia, and music concerts, Yucatecan trova musicians and their patrons celebrate the *Month of the Trovador* with free Yucatecan trova concerts around the city. Since the 1980s *Otoño Cultural* (Cultural Autumn) is a citywide festival that takes place during September and October. It highlights Yucatecan artists at local theaters and venues, usually for free or at low, subsidized prices, and gives free access to all museums and galleries across the city. October is also the month in which the city features the International Festival of Maya Culture (known as FICMaya), which includes conferences, exhibits, concerts featuring local and international artists, and stands selling mostly local food, across the state of Yucatán. *Hanal Pixán* (Day of the Dead) is now a festival with exhibits and flea markets across town, for which the city of Mérida organizes an annual nightly parade called *Paseo de las ánimas* (Parade of the Souls) from downtown to one of the local cemeteries, usually in or around November 1. Méridans love to dress up, so they create costumes of walking skeletons or dress as their favorite

Star Wars or *animé* character to take part in the parade. The Xmatkuil Fair in November and the first days of December is also an important feature of the city's festivals.

In addition to all these festivals for which the city of Mérida offers free transportation and free shows, exhibits and family activities for all, there are also, at least since 2015, several nights each year called "White Nights." At these nights over 200 music shows are offered at venues specifically set up by the municipality, all city galleries are open and entrance to all museums is free. There are also an Annual Animé Fair and an annual Mérida Comic Convention, originally organized by local entrepreneurs but that have become major city festivals, with parades and limited sponsorship from Mérida's municipality. Also, some years the Red Cross or another charity may organize the Walking Zombies Parade, where hundreds of people dress up and put on zombie make-up, and then enjoy a zombie walk along the Paseo de Montejo avenue; or they may organize a *cochinita pibil* festival, in which the famous Yucatecan dish is offered by dozens of cooks in exchange for charity donations. Méridans take advantage of all these festivals, of other minor ones, and of the large availability of restaurants and bars, to get together, enjoy the music and the shows, dine, drink, and engage in merry sociability. So, it is a fact that whether it is to forget one's troubles, to find comfort in the company of others, or to celebrate little and major achievements, in Mérida the party never stops.

Food, music, and sociability

Korsmeyer (1999: 66–67) rejects Bourdieu's ([1979] 1984) notion that taste is merely a class disposition, part of the habitus of people who grew up to expect things in a certain way and to identify some things as finer than others. According to her, Bourdieu's concept of taste flattens the actual importance of literal taste (as in the taste of things being eaten) into a general sociology of taste that does not distinguish food from other forms of art (Korsmeyer 1999: 67). In her view, the taste of things is fundamental to the building of community and it entails aesthetic experiences that relate them to the arts, even if she does not consider food to be an art in the same sense as the fine arts of painting or sculpture are artistic. She ends her book showing the relevance of food to the building of social relations. Taking examples from painting and world literature, she shows that food and art share many things, but taste, and flavor in particular, can build commonality and feelings of belonging. I agree with her general argument, but there is no question that Bourdieu's understanding of *distinction* as sets of practices with which different social sectors engage so as to distance themselves from others, still has explanatory—or at least heuristic—value.

Food, from my point of view as a scholar of organizations and music, does not seem to be too different from other types of aesthetic experience. Like paintings, sculpture, or music, "food" is not a unitary phenomenon that can be analyzed across all kinds of contexts in all cultures, beyond the fact that we all need to eat to stay alive. "Music" is often taken as a single phenomenon, especially because in English and Romance languages it names time-dependent configurations of rhythm and pitch modulation. However, we all have rhythm in our lives, since our hearts keep a regular beat, we walk putting one foot ahead of the other in an even way, and when we speak we use each language's intonation patterns. Thomas Turino (2008: 2) deems *music* a misnomer because in European languages we tend to encompass different kinds of phenomena under that term. At the same time, what we understand as art and as play seem to be universal human phenomena, even if they occupy different places within the social and personal hierarchies in each place (Turino 2008: 2).

Like music, "food" is a misnomer if used as a blanket concept, because it encompasses many different types of phenomena and occasions. Yes, food is something one may take through the mouth and be used up by the body to create life sustenance, but it can also encompass things that are meant for viewing rather than for eating, as well as things one may eat for ritual purposes more than for alimentary ones. The taste of things is not always the main motivation when one consumes food, nor is the color or the nourishment it provides. As Korsmeyer herself explains, in ritual meals there might be "foods" not to be consumed (1999: 138) but rather to be at the table because of their symbolic value. Cakes at Yucatecan weddings are partly for being consumed and partly for décor. Vegetable, ice, cheese, meat, or fruit-made sculptures at many restaurants around the world (Dubai probably being the most famous city for this) are not to be eaten but rather to demonstrate the skill of the carver and suggest that they can prepare food with surgical yet beautiful precision. In Mérida, as elsewhere, food is consumed either in isolation or in groups. It is not the same for a student to reheat in the microwave oven food that she or he extracted out of a can than to share a meal with other people. Food is often featured at the center of a social gathering, which may also include music but, more than anything, will include conversation and the enjoyment of being with others.

If we apply the notion of a participation continuum to middle- and upper-class sociability in chatter and interpersonal contact, we can lay out a Yucatecan participatory continuum of the occasions when it is acceptable to engage in merry group conversation, which in Yucatán tends to result in everyone speaking at the same time and not necessarily taking turns to speak and listen, to occasions in which people are expected to be together but remain separate from each other and in relative silence, as at the concerts of the Yucatán Symphony Orchestra. Shouting in unison can be part of the "conversation" sometimes, such as in social clubs and parties, when everyone joins in cheers for a person or for the motto of their club, or when everyone shouts the lyrics of songs in the disco while dancing. Food and music may or may not accompany Yucatecan sociability at specific times. It is also possible to identify a continuum of relative participatory cooking, at least in Yucatán, from the occasions in which each person contributes one dish to the party, to the occasions in which a single person cooks at home for everyone, to the occasions in which a catering service is hired, to when eating at those particular establishments that are called *fondas* and *taquerías* (eateries and taco places), and finally to being served at fine restaurants featuring auteur's cuisine. The look, the feel, and the overall presentation of the food will differ greatly and so will the prices on the final bill. We see that neither food nor music nor chatter are, then, inextricable companions to sociability, and even if the music of the symphony orchestra can be adequately described as "social" because it is made by the musicians for the public, in Yucatán music is not always accompanied by food. However, they are often expected to come together.

Ayora-Diaz (2010, 2012) has shown how the yearly cycle of food in Mérida is also tied to the cycle of celebrations, and how the repetition of flavors makes food identifiable as "Yucatecan." There are other foods that would not be classified as such by Yucatecans, even if they were a regular part of the yearly menus. These include most of the food served at parties to celebrate rites of passage (baptisms, weddings, anniversaries, graduations, and coming of age celebrations), and much of the food served at bars, restaurants, and fiesta halls. In the past whatever food was not understood as "Yucatecan" was considered "international," but, as Ayora-Diaz has shown (in Ayora-Diaz, Vargas Cetina and Fernández Repetto 2016), the regional diversification of food choices has led to the subclassification of food served locally as "Mexican," "Indian," "Chinese," "Korean," "Brazilian," or ascribable to some other "nationality," or even to regions within nations: "Oaxacan," "from Michoacán," "from Northern Mexico," "TexMex," "Schezuan," or "from the Pampas." Also, food identified before as "international" is now being claimed as "Yucatecan." Two special examples of this are *pastel de tres leches*, a cake usually served at birthday parties, and *sandwichón*, a type of savory cake-sandwich that resembles savory cakes across the Caribbean but in Yucatán and Mexico is considered quintessentially Yucatecan.

The same happens with music. As I have explained elsewhere (Vargas-Cetina 2017), in the past, composers such as Armando Manzanero and Sergio Esquivel were seen as making music in "international styles." However, today they are seen as quintessentially Yucatecan and their songs, in the original versions or in covers, accompany food in restaurants and celebrations where Yucatecan food is served. In terms of the soundscapes that Méridans expect to find as part of their intense sociability, the sounds of the sea and the sounds of the countryside are considered added bonuses to the sounds of music, chatter, and laughter. In Yucatán, social class distinction, in Bourdieu's terms, comes with income differences but also with differences in the range of foods that people in different social classes consume, as well as differences in the music people in different social classes perform and to which they listen (Ayora-Diaz, Vargas Cetina and Fernandez Repetto 2016). Most Méridans can participate in the complex of The Life Delicious I have been describing here, but we all do from our different stations and diversified lifestyles, and this implies that although we all may share at some point the same locations, these do not always intersect. Taste, as preference for some things over others, is not only a set of acquired dispositions but also a sign of purchasing power. In Mérida, when someone says something is "buenísimo" (the superlative form of "good") often this means that one will get more than one would expect for what one pays. Food quality, service, music, ambiance, and price at a restaurant come together in the judgement of that particular place as "bad," "good," or "buenísimo." In the end, most of us in Mérida enjoy The Life Delicious in one way or another, and the economic flows it generates are inimical to the high quality of life for which the city has become famous worldwide.

Enter the unsavory

Jacques Attali (1977) defines as *noise* the sounds that question the clear transmission of signals and challenge the existing social order, especially the established understandings of *music*. In Mérida, The Life Delicious complex might have entered a new phase in 2018, as it is facing two types of noise challenging it: the increasing violence in the neighboring state of Quintana Roo, where many Méridans have businesses, and the protest against audible sounds and music coming out of the restaurants, bars, fiesta halls, churches, and other venues where celebrations take place. The Life Delicious seems to be facing the unsavory.

What today is the state of Quintana Roo, in the Yucatán Peninsula, used to be part of Yucatán, and it only became a separate federal territory at the beginning of the twentieth century and then a Mexican state in the 1970s, when it became a center for national and international tourism. For many decades, it was a place where many Yucatecans conducted businesses, either by relocating there permanently or by opening branches of their existing commercial establishments. By about 2003 criminality appeared, first in the shape of organized mafias charging for protection services and selling drugs, and since 2013, in the form of overt violence including shootings and gruesome executions. Criminals now assault and kill not only members of other organized delinquents but also regular citizens and tourists. The most violent year in Cancún and along the corridor known as the Mayan Riviera was 2017. Newspapers reported hundreds of violent incidents and the violent execution of 326 people (De Mauleón 2018). Yucatán state still maintains its peaceful and festive lifestyle, but the economic drainage of Yucatán's economy resulting from Yucatecans leaving their highly profitable businesses in Quintana Roo will soon start affecting the state's and especially the city of Mérida's lifestyle.

A second, internal source of disruption to Méridans' sociability practices has emerged from Méridans themselves. The perceived safety and the lively cultural life of Mérida have attracted many foreigners to the city. In the 1990s, Méridans had moved to the suburbs abandoning much of Mérida's downtown. Small theaters, galleries, discos, pubs, restaurants, and nightclubs began to open their doors in the area. At the same time, foreigners from other states of Mexico and from abroad began to purchase and invest in the remodeling of downtown colonial houses either to live in them or to open boutique hotels. In short, Mérida's downtown has become a lively area of the city, where outdoor shows, bars, restaurants, pubs, and gourmet markets offer live music and party entertainment all night, most nights, but where an increasing number of people want to use the night to sleep.

Méridans living downtown are now complaining against "sound speakers emitting vibrations that affect the neighboring properties" and what they called "problems of cohabitation" with bars and night clubs that included "urine, vomit, long lines of loud customers on the streets, use of the neighbors' parking spaces and garbage" (Bote Tun 2018). The newly formed association Todos somos Mérida (We All Are Mérida) are now denouncing not only the restaurants and bars at night, but also the loud music and the noise generated by workshops, factories, promotional campaigns of all sorts, fiesta halls, outdoor markets, and live shows (Poo Hurtado 2018). In March and April of 2018 many neighbors hung large signs on their homes with the legend "No more noise! We want to sleep." The hashtags #LaMúsicaNoEsRuido and #NoMatenLaMúsica (#MusicIsNotNoise and #DontKillTheMusic) took over Facebook, Twitter, and Instagram local accounts, as the musicians and their fans launched a counteroffensive, accusing the "foreigners" of threatening the cultural life of the city. Neighbors in other parts of the city also began to protest against the noise. The local association RockCultura held a round table of two musicians, the director of Urban Development of the City of Mérida, the owner of a downtown pub, and a local expert in sound control. They concluded that it is important to create the conditions for the noise not to be a problem in downtown Mérida. However, they asked the neighbors to be tolerant and not destroy the exciting live music festivals that characterize the city. The video was widely distributed via YouTube and Facebook, and was reviewed in the newspaper *La Jornada Maya* (Contreras 2018).

In *The Soundscape of Modernity*, Emily Thompson (2002: 115–168) analyzes the problems surrounding the definition of noise and the efforts made by the city of New York authorities to develop methods and regulations around noise abatement during the first decades of the twentieth century. Much of Thompson's description and many of the accompanying illustrations of questionnaires, petitions, and cartoons resemble what has been happening in Mérida in 2018. As it was the case in New York, the municipal authorities of Mérida and the Yucatán Congress quickly sided with the neighbors of the downtown area and other neighbors expressing their displeasure at the sound levels characterizing the city's everyday life, but especially the loud sounds emanating from spaces for intense sociability. For many years, the city regulations have made it mandatory to restrict the noise in public areas, and therefore the protests were only demanding the application of existing laws. In April 2018, the city of Mérida announced that its new regulations include the end of live music shows thirty minutes before midnight.

In New York, the Noise Abatement Commission dissolved after two years, in 1933, without having solved the problem. It is very likely that the noise abatement commission now forming in Mérida will go the same way, creating areas of relative low-noise around hospitals and schools. However, the sound of the city comes from, as the downtown neighbors already acknowledge, life itself taking its course in what has become a crowded habitat. While the modalities, budgets, and spaces in which celebrations take place are separated by social class, neighborhood, quality of the food, and taste in music, all Méridans cherish the party life. It is about to change to adjust to the new challenges, but it is unlikely that it will end any time soon. Méridans see their social life, which includes food, music, and conversation, as a fundamental part of who they are as humans, as family members, as friends, and as Yucatecans. As long as Méridans continue to enjoy each other and see life as a series of events including food, music, conversation, and laughter, and as long as the violence from other Mexican states continues to be kept off the borders of the state, Méridans will continue to live The Life Delicious in order to remember, forget, and celebrate.

Notes

1 This title makes reference to *The Life Aquatic with Steve Zissou* (2004). In this film, Zissou has built a fantasy world that is threatened by the reality of the world around.

2 Río Lagartos, San Felipe, Las Coloradas, and El Cuyo are the coastal towns where middle-class and wealthy families from eastern Yucatán (especially from Valladolid and Tizimín) have summer properties.

References

Attali, J. (1977), *Noise: The Political Economy of Music*, Cambridge, MA: MIT Press.

Ayora-Diaz, S. I. (2010), "Regionalism and the Institution of the Yucatecan Gastronomic Field," *Food, Culture and Society*, 13 (3): 397–420.

Ayora-Diaz, S. I. (2012), *Foodscapes, Foodfields and Identities in Yucatán*, Amsterdam: CEDLA, New York: Berghahn.

Ayora-Diaz, S. I., G. Vargas Cetina and F. J. Fernández Repetto (2016), *Cocina, música y comunicación. Tecnologías y estética en el Yucatán contemporáneo*, Mérida: Universidad Autónoma de Yucatán.

Bote Tun, A. (2018), "Exigen un alto a bares en el centro," *Diario de Yucatán*, January 30, 2018, www.yucatan.com.mx/merida/exigen-un-alto-a-bares-del-dentro (accessed January 30, 2018).

Bourdieu, P. ([1979] 1984), *Distinction: A Social Critique of the Judgement of Taste*, Cambridge, MA: Harvard University Press.

CONEVAL (Consejo Nacional para la Evaluación Pública del Desarrollo Social) (2018), "Entidades federativas. Yucatán. Estadísticas de pobreza en Yucatán," www.coneval.org.mx/coordinacion/entidades/Yucatan/Paginas/principal.aspx (accessed March 1, 2018).

Contreras, J. M. (2018), "Músicos, funcionarios y empresarios debaten tema de ruido," *La Jornada Maya*, March 21, 2018, www.lajornadamaya.mx/2018-03-21/Musicos–funcionarios-y-empresarios-debaten-tema-de-ruido (accessed March 21, 2018).

De Mauleón, H. (2018), "Una mujer desató la violencia en Quintana Roo," *El Universal*, January 24, 2018, www.eluniversal.com.mx/columna/hector-de-mauleon/nacion/una-mujer-desato-la-violencia-en-quintana-roo (accessed February 15, 2018).

Korsmeyer, C. (1999), *Making Sense of Taste: Food and Philosophy*, Ithaca, NY: Cornell University Press.

The Life Aquatic with Steve Sizzou (2004), [Film] Dir. W. Anderson, Burbank, CA: Touchstone Pictures.

Martín Briceño, E. (2014), *Allí canta el ave: ensayos sobre música yucateca*, Mérida: Gobierno del Estado de Yucatán, Mexico City: Consejo Nacional para la Cultura y las Artes.

Poo Hurtado, F. (2018), "Vecinos del Centro piden apoyo a las autoridades," *Por Esto. Dignidad, Identidad y Soberanía*, March 22, 2018, www.poresto.net/ver_nota.php?zona=yucatan&idSeccion=6&idTitulo=634741 (accessed March 22, 2018).

Thompson, E. (2002), *The Soundscape of Modernity: Architectural Acoustics and the Culture of Listening in America, 1900–1933*, Boston, MA: MIT Press.

Turino, T. (2008), *Music as Social Life: The Politics of Participation*, Chicago: University of Chicago Press.

Vargas-Cetina, G. (2017), *Beautiful Politics of Music: Trova in Yucatan, Mexico*, Tuscaloosa: University of Alabama Press.

8

To Eat *Chapulines* in Oaxaca, Mexico: One Food, Many Flavors

Jeffrey H. Cohen
The Ohio State University

Paulette Kershenovich Schuster
AMILAT, Open University, The Hebrew University of Jerusalem

Introduction

Chapulines (*saltamontes*) are toasted grasshoppers (*Sphenarium pyrgomorphidae*) celebrated as a traditional food in Oaxaca, Mexico. They are consumed in rural households, by urban dwellers in Oaxaca City, and by tourists visiting the region and its restaurants. Chapulines are important in unique ways for each specific group. For rural Oaxacans, chapulines collected in season and from a family's *milpa* (garden plot) are a low-cost snack for between meals, as well as a high-protein readily available food source and meat alternative. Urban Oaxacans enjoy chapulines as a treat. Purchased from street vendors in small paper bags, like popcorn, chapulines are ubiquitous at local events including baseball games, concerts, and theater performances. Chapulines also appear on the menus of Oaxaca city restaurants. Advertised as a local specialty, and a pre-Columbian tradition, chapulines are highly valued by restaurateurs who create their own Oaxacan authenticity and feature chapulines as appetizers and as special main dishes. Tourists visit these restaurants and consume chapulines to "taste the past" or to challenge their own sense of disgust as they eat what they believe are inedible insects. Oaxacans debate how best to serve chapulines. While some rural folks serve them as a side or main dish, city folks maintain they are best served as a crunchy snack. Still others, particularly upscale restaurateurs and foodies, argue that chapulines should not be rolled into a corn tortilla or sold by the bag, but rather included as part of a complete and filling meal that will carry the consumer to the region's indigenous past.

Chapulines do not mean the same for every consumer and they taste different for each group of consumers. For rural Oaxacans, chapulines are a food that has a place in the local diet. They are just to be eaten. People do not spend time commenting on their taste or texture. Urban Oaxacans eat the same chapulines but for a different reason. They consume chapulines as a snack food at special events; they associate chapulines with free time and their own successes as *urbanos* who have left the rural life behind. Describing the crunch and slightly bitter aftertaste as a traditional flavor that comes from the region's past, they associate chapulines most clearly with leisure rather than as a critical food item in the daily diet. Upscale restaurateurs serve chapulines as a food that is part of Oaxaca's history and label them as authentic markers of the state's indigenous heritage. In the process, they capture some of that authenticity and use it to promote the quality of their foods. Nevertheless, chapulines are typically served to tourists visiting the state as "bugs." Their taste is often enhanced with spice and garlic and they are prepared as special plates and unique appetizers. Presented to consumers as a dare, they comment less on flavor and more on the challenge that comes with eating the inedible (Hamerman 2016; Modlik and Johnston 2017). For tourists chapulines become a challenge that must be met and often conquered.

This chapter explores how chapulines have become distinctive and associated with specific groups—rural Oaxacans, urban Oaxacans, as well as upscale restaurateurs and the tourists who visit their restaurants. While a rural Oaxacan might say chapulines are a readily available food and something that their grandparents ate in moments of crisis, restaurateurs describe them as an example of a pre-Columbian food tradition (on tortillas see Pilcher 2008; Vizcarra Bordi 2006). Understanding the role of chapulines at the Oaxacan table and for rural families, urban Oaxacans, and tourists is an opportunity to explore how tastes shift and how the role and meaning of food changes (see Garibay-Orijel et al. 2007; Sammells 1998). After briefly introducing what chapulines are and how they are prepared, we focus on how chapulines taste for rural Oaxacans, city folk, as well as restaurateurs and tourists. We argue that chapulines are best understood not as a single food but instead as a collective food that has different meanings and tastes for different people. Our analysis captures the differences in use and taste as we define the shifting role and value chapulines hold in the Oaxacan diet.

Notes on production

Grasshoppers are found throughout the central valleys of Oaxaca and while they can be collected almost anywhere, most grasshoppers that will become chapulines are collected by hand and from a family's milpa. Prepared chapulines are found year round in stores and on menus; however, they are a seasonal delicacy and most often served freshly prepared. The chapulines season begins with the arrival of the spring rains in April and continues through the late fall. Cooler temperatures in September and October tend to signal the end of the season and a natural decline in the population (Ramos-Elorduy et al. 1997). Grasshoppers emerge from eggs and through a process of incomplete metamorphosis, reach adulthood in four to seven weeks. Young grasshoppers or nymphs (often called babies by producers) are valued over adults, and can sell for double the price of fully grown, prepared chapulines (Cerritos Flore, Ponce-Reyes and Rojas-García 2015).

Production is not complicated but it is time consuming and laborious for the *chapulineras* involved (see Cerritos and Cano-Santana 2008).[1] Once the season has begun, women will travel to their family's milpa early in the morning, before the grasshoppers have time to warm up. Cool, inactive, and easier to catch, chapulineras and their children will walk through their milpa, waving nets and bags to capture the grasshoppers. Once caught, the grasshoppers are then transferred to closed boxes and placed in cool corners of rooms at home or on the patio. There they relax and when calm they are sorted by the chapulineras who organize them by size and dispose of any other insects (for a full description of production see Cohen, Mata Sánchez and Montiel-Ishino 2009).

Preparing chapulines is not difficult but, again, it can be time consuming. The chapulinera brings water to a boil and seasons it with a little garlic and lemon. Then she dunks the grasshoppers in the boiling water to kill them, similar to how lobsters are prepared. The chapulines are next transferred to a *comal* (a clay or cast iron griddle) for toasting until they gain their red color.[2] Freshly prepared chapulines do not have a long shelf life and they can begin to spoil in just a few days. Generally, chapulineras will prepare only enough chapulines to sell each day, leaving boxes of grasshoppers to sit in a cool room until they are needed. Once prepared, chapulines are sold in markets throughout Oaxaca City. Bought by weight, chapulines are expensive and often a kilogram costs as much as meat. Over the last decade, the price of chapulines in Oaxaca's markets has hovered to around 100 pesos a kilo. In the early part of the twenty-first century this price translated to about US$10 per kilo or just under US$5 a pound. Of course, chapulineras have access to chapulines directly. Nevertheless, our surveys of eating habits conducted in 2006, 2007, and 2008 revealed that even the few rural Oaxacan families that ate chapulines did not rely on them as a regular part of their daily diet.

Chapulines are not madeleines (what rural Oaxacans eat)

Marcel Proust describes savoring the moment as he dips a madeleine in tea and feels a rush as he is overwhelmed by nostalgia for the joys of his childhood in his book, *Remembrance of Things Past*. For him, and for many of us, the act of eating brings with it "the vast structure of recollection" (Proust 1992: 66; and see Sutton 2001). But is Proust's recollections of the past a constructed memory of a time that never was? The aroma of the madeleine carries Proust back to his childhood, but it is an imagined childhood and a selective memory. The fears of war and the threat of violence that surrounded Proust and his family fade in his telling. They are replaced with memories of experiences free of trouble, captured in a cookie. Proust is secure in the past that is created by the aroma of the madeleines and that nostalgically evoke memories. He is surrounded by love and welcomed by support even as the violence of the early twentieth century swirls around him just offstage.

We might assume that chapulines, like the madeleines for Proust, serve much the same role and carry the consumer or eater back to a bucolic past and create a sense of home and control (see Dudley 2011). In fact, research into the nostalgia that surrounds tortillas suggests that simple foods and their aromas can carry a consumer back to their past. Furthermore, Vizcarra Bordi (2006), writing on the important role tortillas can play in the lives of Mexicans, notes how consumers use tortillas to challenge the globalizing forces of Western style foods and eating. But chapulines are not madeleines for rural Oaxacans. When rural Oaxacans consume chapulines they are not overcome with a positive, warm, and welcoming nostalgia for the past. The smell, taste, and presentation of chapulines do not establish home and erase uncertainty. Rather, chapulines carry an ambivalence that is bittersweet and worrisome at best (Vignolles and Pichon 2014: 228). When Cohen asked older Oaxacans to comment on eating in the past, whether they talked about chapulines, tortillas, or something else, they typically responded: why go backwards? Who would want to return to the deprivation, poverty, and uncertainty of the past?

When Cohen sat with don Librado in a bare, dirt floor adobe room in the early 1990s, don Librado talked about the past and the foods he remembered eating. He was born in the 1930s and shared his childhood and the stories he had heard from his parents about the Mexican Revolution. He recalled how his parents and other villagers were forced to abandon their homes while factions of at least two armies fought for control of Oaxaca City (Cohen 1999: 30–32). During much of the revolution, *santañeros* (villagers from Santa Ana) sought refuge in Las Carretas, a mountain settlement perched above the village proper. Most of the villagers who fled the onslaught lost access to their fields and gardens. Food was hard to come by. Don Librado recalled that his parents were always hungry. He described how they turned to wild plants, wild game, chapulines, and other insects. Villagers remained in Las Carretas for nearly three years, later returning to the valley to rebuild homes and replant milpas.

Working with don Librado and other older villagers, Cohen pieced together some of the events and experiences that characterized their time in Las Carretas. Chapulines were a central food. They were eaten with tortillas and represented an important source of protein in the diet. But chapulines were not beloved, and they were never more than something to eat. Chapulines did not hold special meaning and were not a traditional value that rural Oaxacans felt they must eat; instead they were food and little more. There was no nostalgia for life in Las Carretas, no urge to memorialize the suffering that came with that life or celebrate the foods that kept people alive. Chapulines filled an important role and took the place of foods people missed. People liked them, but they were not celebrated as a rare or tasty delicacy. Chapulines were not a food that people wanted to remember, but rather chapulines were a choice of last resort, a disaster food and a means for survival. Memories of chapulines were of a food that served an important purpose when there were few other alternatives (Sutton 2001). Memory was strong among most santañeros Cohen talked to. They had a small community museum that commemorated the past and documented the years of the Mexican Revolution among other things (Cohen 2001). Don Librado and others took great pride in celebrating village heritage and their community's resilience and strength. When Cohen asked about returning to that world, or more precisely to the world before the disruptions of the revolution, villagers were blunt and emphatic, "no." Doña Anita was clear, "We had to eat grass. Grass! It wasn't easy. It made no sense. No, I would never go back there." It is clear that chapulines played a role as a disaster or crisis food for most rural Oaxacans. And while they are celebrated in upscale restaurants, their role in the rural household and through most of the twentieth century is not well documented (Endfield, Fernández Tejedo and O'Hara 2004).

During Cohen's fieldwork, when Cohen and his students were asking specific questions about food, farming, and selling chapulines, people did not talk about them. While there are chapulineras who will harvest, prepare, and sell chapulines in regional markets throughout the central valleys, few people talked about eating them as part of an everyday diet. They were described as food, with no special meaning. The rather regular taste and value of chapulines extended to their preparation. When Oaxacans talked about cooking and serving chapulines, they shared only basic recipes. Typically, santañeros prepared chapulines simply or bought them in the market, and ate them rolled into a tortilla. Don Mauro, like many of the men Cohen interviewed in the 1990s, often carried a bag of chapulines with him to snack on while going about his daily business. But chapulines never showed up on his plate as part of a main meal. And while rural Oaxacans and nearly everyone Cohen interviewed could describe the flavor and aroma of homemade tortillas, chapulines were just chapulines. And that was how most people ate chapulines, whether gathering firewood, moving livestock, farming in the early morning, or weaving, santañeros would snack on chapulines. But snacking on chapulines was a way to put off hunger; it was not a nostalgic or melancholic act.

One of the first experiences came as Cohen accompanied don Mauro to collect firewood. It was a beautiful cool morning, and the sun was just rising over the valley. They drove up to the mountains, and found a scrub forest filled with songbirds. After about an hour, they took a break. don Mauro pulled out a bag of chapulines and a small stack of tortillas. Sitting on a tree stump, with the valley spread out below and the sun warming them, don Mauro gave Cohen a chapulines eating lesson. With a smile, he handed him a large tortilla. He told Cohen to hold it open and poured on the chapulines. He did the same with a second tortilla for himself. Gesturing that Cohen follow along, he rolled up the tortilla, making a taco. He took a bite and knowing that Cohen was not so enthusiastic, he encouraged him to join him. Embracing anthropological traditions and the rather dated saying "when in Rome" he took his first bite. Not at all "buggy," but a bit pungent and bitter, he thought. Rolled into a tortilla, the chapulines were easy to eat. They were a little crunchy. Don Mauro warned him to pull the legs off or to spit them out, saying "They will scratch your throat if you swallow them." That he had gotten the *gringo* (a popular appellative for Americans) to try chapulines was not lost on don Mauro as they continued their day. They spent another hour collecting firewood before returning home for what he described as a "proper" breakfast that included farm fresh eggs, black beans, tortillas, and *salsa*, as well as strong coffee. In contrast, Schuster, a native Mexican, never ate chapulines while living in Mexico. She ate her first chapulin in Israel of all places. One of her students had recently visited Mexico and offered her some of the "goodies" he had brought back. Curious as to what he meant, the student pulled out a plastic container full of chapulines. Surprisingly, the student enjoyed them, even citing that they were kosher (following Jewish dietary law) as he handed the container to Schuster. Schuster promptly ate one without really thinking of the act of consumption. She ate it as a means to fulfill a curiosity, not a wanting need or as part of a thrill seeking adventure.

Maybe chapulines are madeleines (what urban Oaxacans eat)

In contrast to rural Oaxacans, urban Oaxacans (folks who have settled in Oaxaca city) who consume chapulines, talk about them in terms that are celebratory and steeped in a positive nostalgia (Korsmeyer and Sutton 2011). While the past these consumers evoke likely never existed, chapulines are celebrated as a connection to the state's indigenous heritage. Chapulines are an important seasonal snack food. Eaten like popcorn, they are a slightly spicy treat and a regular feature at Oaxaca's Mexican league baseball games and at market stalls.

Other Oaxacans incorporate chapulines into their meals that celebrate and preserve what some city folks describe as the rustic flavors of rural life and a lost past. Restaurateurs build upon this idea and present chapulines as delicacies that bring the foods and flavors of the state's indigenous past into the present (Heldke 2015). They also use the chapulines to wrap their restaurants in authenticity and create what Grasseni (2005: 93) defines as *different timescapes* or *temporalities*. These different or alternative timescapes interpret rural agriculture for an urban audience and the tourists who visit the region, and present that agriculture as a remnant of Oaxaca's indigenous past. Whether those traditions are "real" or "invented" or "imagined" and part of a process of ongoing mythologizing Oaxaca's past (Goertzen 2010) is not important. Businessmen and women use these images to sell their goods and create a sense of continuity between past and present that is manifest in the foods, drinks, and services they sell.

There is one additional way of consuming chapulines that is not based in Oaxacan traditions or practices but instead serves tourists. For the tourists, eating chapulines is a challenge. Chapulines "dare" the tourist to consume something that they believe is inedible, that it isn't food but a strange thing. Eating a bug on a dare allows the tourist to conquer the unknown while they "eat like an Indian." This act feeds into present race relations and past mythological cleavages. While rural Oaxacans consume chapulines and taste food, urban Oaxacans taste leisure. The value and role of chapulines shift for the restaurateur who may emphasize their meaning over flavor and as they use chapulines to evoke an authentic connection to Oaxaca's past. Finally, the tourists misread the taste and value of chapulines as they focus on the dare that is involved with eating the inedible. For the tourists taste is defined around the fact that they are eating an insect, not necessarily by the insect's flavor.

Celebrating the past

At food street stands and at special events, particularly sporting events, chapulines are sold in small paper bags alongside popcorn and other snack foods.[3] Urban, city dwellers described eating chapulines as fun and as a treat that signals leisure time, happiness, and celebration. Robin Fox, writing on the "myth of nutrition," argues that people eat to feel good and to celebrate (1994). The smell and taste of chapulines helps transform the everyday and suggests it is time to enjoy. City workers described carrying a small bag of chapulines to their offices and sharing them with friends during breaks. Everyone recognizes the pungent aroma of the chapulines and pauses in their routines to share.

Thome-Ortiz and Moctezuma-Pérez (2015: 73) argue that the popularity of certain foods creates an experience that evokes a sense of pleasure that can be both physical and psychological. Restaurants play upon this process, as well as the idea that chapulines were common in Oaxaca's indigenous past. They serve chapulines in a few different ways. They prepare them as appetizers and as parts of happy hours, and use them to symbolize and celebrate Oaxacan traditions. More informally, restaurateurs and their waiting staff will challenge visiting eaters into trying something authentic (an insect), a food that typically sits well outside what is considered "good." The connection between past and present, as well as psychological and physical well-being, is emphasized as chapulines become a main dish in urban households (Sutton 2010).

From the technology of cooking to the aroma, taste, and dish organization, chapulines carry urban Oaxacans away from the city and to a rural and bucolic past, and what they assume is a more peaceful lifestyle. At homes, where chapulines are collected, they are prepared and presented as something special. While everyday cooking encourages quick and filling meals built around basic starches (particularly bread, pasta, rice, and tortillas) and the use of prepared ingredients that are easy to transform, preparing chapulines focuses on what are described as traditional food themes—tortillas are present and meat (beef, chicken, or pork) may be included, but the key feature are the chapulines.

Preparation can take many forms, including *salsas* (hot sauces) and *cremas* (cream sauces). Dishes include chapulines vinaigrette, chapulines hash, and chapulines with mixed vegetables, in addition to long-established dishes such as chapulines tacos, *chiles rellenos* (stuffed chili peppers) of chapulines, and chapulines *estofado* (stew).[4]

In nearly every case, the preparation of the meal comes with a discussion of chapulines, the assumed traditions that surround their use, and how the dish carries its maker and eater to an earlier, more rural, and bucolic/ideal time. And while the restaurant and waiting staff will celebrate what they see as the importance of chapulines in the history of rural Oaxaca and indigenous culture, they will sometimes also emphasize the sustainability of insects and the ways that entomophagy can enhance health. Playing with the way the taste of chapulines is interpreted differently by different consumers, restaurateurs can be critical of rural Oaxacans as they describe how they preserved the dynamic ways in which chapulines are prepared and noting that rural families have lost the ability to make anything other than simple chapulines tacos.

Eating bugs

Most tourists to Oaxaca are members of Mexico's middle and upper classes, or foreigners who are most often from the United States. The Mexican Secretary of Tourism (SECTUR) notes that in 2014, over three million tourists visited the state. Of this group, 2.95 million were native to Mexico and just over 131,000 came from abroad (SECTUR 2014). These are tourists who want to eat and celebrate, but they also want good food. And while entomophagy is common throughout most of Mexico, its prevalence has been in decline as incomes rise and Western style diets (centered around meat) grow to dominate the nation (Van Huis et al. 2013). Thus, while there are hundreds of restaurants in Oaxaca, and many that specialize in traditional and typical foods, including chapulines, demand is often framed around the dare rather than tradition. Meeting the demands of tourists, restaurants and cafes present chapulines as a local delicacy and also as a healthy dare, "can you eat a bug?" emphasizing both the challenge of eating something that is assumed to be disgusting, as well as rich in nutritional value. Chapulines are low in fat and rich in amino acids and easily digestible protein (Ladrón de Guevara et al. 1995; Ramos-Elorduy et al. 1997). Emphasizing chapulines' nutritional value, particularly for weight-conscious tourists, helps restaurateurs and their waiting staff to entice tourists to try eating bugs. Touted as inexpensive, sustainable, and environmentally friendly food, tourists are dared to eat them. And many do. Nevertheless, chapulines remain a minor player on most menus that include other local traditions such as *moles, tlayudas, memelas,* and *tasajo* (Kennedy 2010).

Discussion

Chapulines have distinctive tastes and fulfill unique roles for different groups of individuals. Most rural Oaxacans do not think a lot about chapulines. As a food, chapulines are defined not by a special taste or by some deeply held traditional value, but because they are something to eat. Most rural folks think of chapulines as a cheap, alternative food. They might classify them as something that their parents or grandparents depended on in times of hunger, whether caused by natural disasters or the violence of warfare. City folks living in the state's capital often snack on chapulines. They associate chapulines with the rural past of the state that has disappeared and, more importantly, with fun and leisure. Found at sporting events, chapulines now represent a special food, not something to be consumed regularly but something to be enjoyed as part of a day off. Restaurateurs carry the meaning of chapulines in a different direction, as they use chapulines to promote the authenticity of their food—it is traditional Oaxacan fare—and in the process build their business. A special flavor, one that harkens back to the state's past, also celebrates the efforts of the entrepreneur to document that past.

Tourists enjoy chapulines as part of Mexico's past and find excitement in them as a "dare." Tourists do not appear to enjoy chapulines as a delicacy with a special and unique taste. Instead, chapulines are inedible bugs and eating them shows the courage and strength of the consumer. Ignoring the chapulines' taste and focusing on the dare might seem surprising given the many ways that they are carefully prepared and presented by restaurateurs; nevertheless, tourists to the region (including foodies) seem more interested in the inedible qualities of chapulines as insects rather than their taste.

Chapulines are not like tortillas, as they are not a central feature of the meal. And consumers, whether at the rural table or the fine restaurant, are not looking for them on the table. Nevertheless, when they do appear, they hold important meaning. We believe that approaching taste as a complex site that changes in relation to use and meaning can help us understand the different ways that chapulines are consumed, and how they are used by rural Oaxacans, city folks, and restaurateurs, as well as by tourists/foodies. Expanding how we talk about chapulines allows our discussion of culture, food, and consumption to move beyond what our informants put in their mouths and develops a dynamic understanding of the way taste connects to culture and a changing society (Sutton 2010).

For rural Oaxacans, chapulines are a disaster/survival food and their taste is secondary. Experts on entomophagy describe chapulines as a pre-Colombian resource; nevertheless, unlike insect derived foods that we know about (Van Itterbeeck and Van Huis 2012) there is little evidence that chapulines were more than a crisis food in times of drought (Endfield, Fernández Tejedo and O'Hara 2004). Urban Oaxacans use and consume chapulines in a way that is different from their rural counterparts. Chapulines are a nostalgic snack food. Like Proust's madeleines, chapulines conjure up a rural past that no longer exists (if it ever did). They signal leisure time, and for many city folks their new patterns of consumption and taste show how different they are from the rural folks who may live in the towns that were their childhood homes. Restaurateurs take a different tack as they approach chapulines. They develop creative and innovative dishes around the grasshoppers, and typically try to link those dishes with the past. They celebrate Oaxaca's heritage, creating dishes that highlight tastes, commitment, and define their restaurant's pedigree (it is really Oaxacan, not some transplant). Finally, for the visitor to the region who is not moved to "eat like a native" or "taste" the flavors of history, chapulines are a healthy dare. They include a great deal of nutrition in a small package and for some tourists that is enough. They will give them a try; and while they may be put off by the idea of eating a bug, they are willing to show off and show their strength by eating what should be inedible.

Taste has a biological dimension. Taste receptors respond to tastants and signal if something is sweet, salty, sour, bitter, or *umami*. But most importantly, taste is also learned and cultural. For tourists, chapulines are disgusting, not because they taste bad but because they are assumed to be vile. Rural Oaxacans eat chapulines and enjoy a toasted grasshopper. These chapulines are somewhat chewy, slightly bitter and a little spicy with a strong aftertaste, but we can all learn to eat chapulines. Rural Oaxacans who sought alternative foods during the Mexican Revolution turned to chapulines, and most eaters find that their tastes change over time as well. That our tastes can change complicates our ability to understand what tastes "good" and what does not for any given people at any given time. By analyzing how chapulines are markers of social stratification and difference for different groups we can better understand how people in

Oaxaca—both tourists and local—classify foods, flavors, and tastes, what their different appreciations mean, and how those classifications change locally and globally in the tourism market.

Notes

1 While there is no rule that defines producers as women, chapulineras are typically women and production is described as a woman's job.

2 The bond between the protein astaxanthin and the natural pigments in the grasshoppers breaks down in response to heat allowing the natural color, a bright red, to show. Lobsters turn red in much the same way.

3 Chapulines appeared at Mariners' games in Seattle, Washington, in much the same role.

4 There are many businesses in the United States and Europe that produce grasshopper or cricket flours. These are not served in Oaxacan restaurants.

References

Cerritos Flore, R., R. Ponce-Reyes and F. Rojas-García (2015), "Exploiting a Pest Insect Species Sphenarium purpurascens for Human Consumption: Ecological, Social, and Economic Repercussions," *Journal of Insects as Food and Feed*, 1 (1): 75–84.

Cerritos, R. and Z. Cano-Santana (2008), "Harvesting Grasshoppers Sphenarium purpurascens in Mexico for Human Consumption: A Comparison with Insecticidal Control for Managing Pest Outbreaks," *Crop Protection*, 27 (3): 473–480.

Cohen, J. H. (1999), *Cooperation and Community: Economy and Society in Oaxaca*, Austin: University of Texas Press.

Cohen, J. H. (2001), "The Shan-Dany Museum: Community, Economics and Cultural Traditions in a Rural Mexican Village," *Human Organization*, 60 (3): 272–280.

Cohen, J. H, N. D. Mata Sánchez and F. Montiel-Ishino (2009), "Chapulines and Food Choices in Rural Oaxaca," *Gastronomica*, 9 (1): 61–65.

Dudley, S. (2011), "Feeling at Home: Producing and Consuming Things in Karenni Refugee Camps on the Thai-Burma Border," *Population, Space and Place*, 17 (6): 742–755.

Endfield, G. H., I. Fernández Tejedo and S. L. O'Hara (2004), "Drought and Disputes, Deluge and Dearth: Climatic Variability and Human Response in Colonial Oaxaca, Mexico," *Journal of Historical Geography*, 30 (2): 249–276.

Fox, R. (1994), *The Challenge of Anthropology: Old Encounters and New Excursions*, New Brunswick, NJ: Transactions.

Garibay-Orijel, R., A. Estrada-Torres, J. Cifuentes and J. Caballero (2007), "Understanding Cultural Significance, the Edible Mushrooms Case," *Journal of Ethnobiology and Ethnomedicine*, 3 (1): 1–18.

Goertzen, C. (2010), *Made in Mexico: Tradition, Tourism, and Political Ferment in Oaxaca*, Jackson: University Press of Mississippi.

Grasseni, C. (2005), "Slow Food, Fast Genes: Timescapes of Authenticity and Innovation in the Anthropology of Food," *Cambridge Anthropology*, 25 (2): 79–94.

Hamerman, E. J. (2016), "Cooking and Disgust Sensitivity Influence Preference for Attending Insect-Based Food Events," *Appetite*, 96: 319–326.

Heldke, L. (2015), *Exotic Appetite: Ruminations of a Food Adventurer*, New York: Routledge.

Kennedy, D. (2010), *Oaxaca al gusto: An Infinite Gastronomy*, Austin: University of Texas Press.

Korsmeyer, C. and D. Sutton (2011), "The Sensory Experience of Food," *Food, Culture and Society*, 14 (4): 461–475.

Ladrón de Guevara, O., P. Padilla, L. García, J. M. Pino and J. Ramos-Elorduy (1995), "Amino Acid Determination in Some Edible Mexican Insects," *Amino Acids*, 9 (2): 161–173.

Modlik, M. and L. Johnston (2017), "Huhu Grubs, Bull Semen Shots and Koki: Visceral Geographies of Regional Food Festivals in Aotearoa," *New Zealand Geographer*, 73 (1): 25–34.

Pilcher, J. (2008), "The Globalization of Mexican Cuisine," *History Compass*, 6 (2): 529–551.

Proust, M. (1992), *In Search of Lost Time, Volume 1 Swann's Way*, revised by D. J. Enright, trans. C. K. Scott Moncrief and T. Kilmartin, New York: The Modern Library.

Ramos-Elorduy, J., J. M. Pino Moreno, E. Escamilla Prado, M. Alvarado Perez, J. Lagunez Otero and O. Ladrón de Guevara (1997), "Nutritional Value of Edible Insects from the State of Oaxaca," *Journal of Food Composition and Analysis*, 10 (2): 142–157.

Sammells, C. A. (1998), "Folklore, Food, and National Identity: Urban Legends of Llama Meat in La Paz, Bolivia," *Journal of the International Society for Contemporary Legend Research*, n.s., 1: 21–54.

SECTUR (Mexican Secretary of Tourism) (2014), *Compendio Estadístico del Turismo en México Mexico*, Secretaria de Turismo, Mexico City: SECTUR.

Sutton, D. E. (2001), *Remembrance of Repasts: An Anthropology of Food and Memory*, Oxford: Berg.

Sutton, D. E. (2010), "Food and the Senses," *Annual Review of Anthropology*, 39 (1): 209–223.

Thome-Ortiz, H. and S. Moctezuma-Pérez (2015), "Leisure Time and Food Memory: An Anthropological Approach to Culinary Tourism in Central Mexico," *Folia Turistica*, 37: 69–83.

Van Huis, A., J. van Itterbeeck, H. Klunder, E. Mertens, A. Halloran, G. Muir and P. Vantomme (2013), *Edible Insects: Future Prospects for Food and Feed Security*, Rome: Food and Agriculture Organization of the United Nations.

Van Itterbeeck, J. and A. van Huis (2012), "Environmental Manipulation for Edible Insect Procurement: A Historical Perspective," *Journal of Ethnobiology and Ethnomedicine*, 8 (1): 1–7.

Vignolles, A. and P.-E. Pichon (2014), "A Taste of Nostalgia," *Qualitative Market Research: An International Journal*, 17 (3): 225–238.

Vizcarra Bordi, I. (2006), "The 'Authentic' Taco and Peasant Women: Nostalgic Consumption in the Era of Globalization," *Culture & Agriculture*, 28 (2): 97–107.

Part Three

Taste and Displacement: Mexican Food in the World of Consumption

9

The Taste of Oaxaca: It's to Die For!

Ramona L. Pérez
San Diego State University

Introduction: The unique flavors of Oaxaca

Oaxaca is a culinary mecca that has become known for the breadth of its regional cuisine.[1] It is a tourist destination where gusto is not just a culinary phenomenon but also a multisensory process of "being" in a land that many consider a fusion of precontact, colonial, and contemporary cultures. The state of Oaxaca has long been known for its multitude of archaeological sites and its lively folk art but its cuisine was a quiet gem that only those who came and indulged knew. Today, people travel to Oaxaca specifically for its cuisine, seeking the unique tastes that can only be found there.

The tourist-based food explosion in Oaxaca occurred over the last several years as a result of Mexico's campaign to take advantage of UNESCO's inclusion of Mexican traditional cuisine on its 2010 "Humanity's Intangible Cultural Heritage" list. UNESCO recognized that, "Traditional Mexican cuisine is central to the cultural identity of the communities that practice and transmit it from generation to generation" (UNESCO 2010: 41). They go on to note that the recognition is not just about the cuisine but also the full cycle of food production, "comprising farming, ritual practices, age-old skills, culinary techniques and ancestral community customs and manners" (UNESCO 2010). While the food of Mexico is steeped in the indigenous foods of corn, beans, squash, chilies, tomatoes, cocoa, avocado, and a plethora of herbs, it is made regionally unique through the integration of local foodstuffs, cooking technologies, and the culinary influences of various external populations over time (see Ayora-Diaz, this volume). In Oaxaca, the cuisine reflects its unique climatic environments, cultural diversity, pre- and postcolonial history, long-held culinary practices, and, more recently, contemporary engagements in the global tourist industry.

It didn't take long following the UNESCO designation for the food-centric tourist media to pick up Oaxaca and cart the state and its local *cocineras* into the international spotlight. *Lonely Planet* named Oaxaca "the gourmet heart of Mexico" (Gilbert 2014) and other online sites call out Oaxaca as the new mecca for food, challenging Mexico City (Cocking 2017). Anya von Bremzen (2016), writing for *Travel and Leisure,* argues that Oaxaca "has always been Mexico's most authentic food town" but that the return of native chefs from restaurants in Europe and the United States has brought a newfound appreciation for the roughly 500 edible herbs that make their way into traditional cuisine. Rubenstein, in her article for *The Atlantic* in 2012, focused away from the growth of new chefs and highlighted the wide range of foodstuffs that the locals enjoy in regional marketplaces. Indeed, Oaxaca's cuisine remains grounded in long-standing ingredients and practices that shape its unique *sabor* and that continues to draw displaced Oaxaqueños back home. At the turn of the twenty-first century, the pursuit of gusto through global food tourism is clearly a booming market that reifies local identities and affirms traditional practices and tastes (Baldacchino 2015; Everett and Aitchison 2008; Holtzman 2006; Lin, Perason and Cai 2011).

Oaxaca's cuisine reflects their distinctive history, landscape, and climate. The state has a plethora of microclimates that range from the mangrove forests of the Pacific Ocean to the deciduous scrublands of the Tehuacán desert on its northern border with the state of Puebla, and with altitudes ranging from sea level to more than 3,500 meters (Lorence and Mendoza 1989). With five major rivers running from the state's three mountain ranges into the Pacific Ocean, the people of Oaxaca have been able to cultivate and forage an enormous range of foodstuffs that have resulted in a broad range of tastes, textures, and smells that one encounters throughout the state. The range of microclimates that fostered the diverse native diet was also instrumental to the integration of flora and fauna that were introduced during the colonial and postcolonial periods. Citrus, coffee, peach, plum, and apple trees dot the landscape of Oaxaca today and enhance the native fruit basket of melons, bananas, guanabana, mamey, sapodilla, maracuja, papaya, mango, cherimoya, and the fruit of the cactus. Oaxaca's primary agricultural exports include limes, mangos, coffee, peanuts, sugar cane, and, as Brulotte (this volume) demonstrates, maguey. The ancient cuisine of Oaxaca was primarily vegetarian with protein drawn from legumes, nuts, and seeds. Meat was served sparingly and included peccary, tepezcuintle, deer, and a variety of birds and other small animals. The introduction of cattle, sheep, pigs, and goats have greatly modified the Oaxacan diet over time, adding not only meat to dishes but also the development of cheeses and milk products for which Oaxaca has become famous.

The most famous of Oaxaca's cuisine, however, lies in its velvety *moles*, which embrace the precontact ingredients of dried chili peppers, ground corn, plantains, chocolate, ground pumpkin or squash seeds, and herbs such as epazote (*dysphania ambrosioides*), pitiona (*lippie alba*), cloves, and chípil, also known as chepil, or chipilín (*crotalaria longirostrata*), along with foodstuffs introduced by the Europeans such as dried plums, peaches, pineapple, almonds, sesame seeds, and cinnamon. The extensive list of ingredients, intense labor, and time necessary to prepare them, let guests know that whenever moles are served, they have come to a special celebration. And while the various moles are comprised of anywhere from a dozen to more than two dozen ingredients, it is the base ingredient of chili peppers that defines most moles. *Mole negro* (black mole), the quintessential mole of Oaxaca, is made with six different chili peppers but it is the *chilhuacle negro* that makes it unique. *Mole Amarillo,* another special mole of Oaxaca, is made with the *chilhuacle amarillo*; both varieties, along with the *chilhuacle rojo*, are grown in the La Cañada region of Oaxaca. While chilhuacles can be purchased in other parts of Mexico, the unique taste that is drawn from the environment of the La Cañada region and embedded in these chili peppers is so distinctive that it demands a high price; Oaxacan-grown chilhuacles can be purchased online for a price that at the time of writing runs at US$75 per pound and roughly US$25 per pound if purchased while in Oaxaca.

Equally significant to the daily palate of local people are the savory *caldos* or clear broths that fuse local traditional plants such as *quelites* (young, tender leaves) of cenizo, chilacayota, chayote, amaranto, or any number of other plants, with edible herbs. Although traditionally served without meat, today they usually include chicken, turkey, or beef. The caldos are served with fresh corn tortillas or dumplings made with ground corn and herbs. For me, no trip to Oaxaca is complete until I have had a steaming bowl of *sopa de guias,* a broth made from the tendrils and young leaves of the squash plant, sliced corn still on the cob, squash, onion, garlic, squash flowers, and dumplings made of corn masa and chepíl. Even salsas have a unique taste that is derived through the choice of chili peppers, tomatoes, tomatillos, salt, and herbs from the precontact period that are then melded with onion, garlic, and sometimes limes, mango, or other fruits. Seasonal delights such as *chapulines* (grasshoppers) or *gusanos* (moth larvae) are added for both taste and increased nutritional value. For a thicker, paste-like salsa with a nice smoky taste, *chicatanas* (flying ants) are ground with salt and chili peppers, and spread on a tortilla.

And yet, even if one opted not to add chapulines, gusanos, or chicatanas to their recipes, one cannot reproduce the taste of Oaxaca outside of Oaxaca—ask any tourist who has collected recipes and then tried to recreate their favorite Oaxacan dishes at home. Oaxaqueños living in the United States also acknowledge that while they can obtain nearly all of the ingredients for their favorite dishes in the United States or even other parts of Mexico, they just don't taste the same (Pérez, Handley and Grieshop 2010). Instead, when they head back to Oaxaca for special holidays they return with suitcases full of local ingredients to make their favorite dishes, and when they cannot travel back to Oaxaca their families collect, toast, and pack the ingredients to have them flown directly to them (Grieshop 2006; Pérez, Handley and Grieshop 2010). It is in this way that they can fill their hearts with home again, an issue I describe in more detail below.

It is not just the ingredients that produce the sabor of Oaxaca and provide the gusto tourists and displaced locals seek, it is also the processes of the *cocinera* and the tools of the kitchen. In many ways, it is here, in the human interaction with the foodstuffs gathered from the environment, that food for satiation transforms into gusto, food for pleasure, and food as identity.

Adding sabor through process

My father used to refuse to eat salsa or guacamole that wasn't prepared in my mother's *molcajete*, a granite stone mortar and pestle, claiming that it had no taste if it was made any other way. Although we tried many times to hide the use of the blender, our apparatus of choice, by serving the salsa in the molcajete, stirring it around a bit to gather up the taste of the stone, he always seemed to know. Like the soil, water, and air that form the tastes in the chili peppers, vegetables, and herbs we use, the practices associated with food preparation create taste and secure our identities to the foods we eat. Taste and the practices associated with its production, as Bourdieu ([1979] 1984) noted, form a part of our identity that defines our membership in our culture group. Sutton, in reviewing the work of Mary Douglas and other anthropologists who studied food as part of ethnicity, argues that food "is a particularly good 'boundary marker', perhaps because it provides a potent symbol of the ability to transform the outside into the inside" (2001: 5). What has stood out for me over the last twenty years that I have cooked alongside Oaxacan cocineras, is their adherence to some traditions over others in the process of preparing dishes for their families and their communities. For instance, while many women in urban and peri-urban spaces have become accustomed to taking their *nixtamal* (dried corn that has been soaked overnight in a mixture of water and lime) or mole ingredients to a local *molinero* for grinding in lieu of using the traditional *metate*, few will give up their ceramic *comales, cazuelas*, or *ollas* (griddles, casseroles, and pots) that are used to heat, simmer, steam, or stew their precious ingredients, turning them into the dishes that their families crave. Women's ability to reproduce the recipes of their community is only one aspect of how food plays into a community's identity. As Bessière has argued, "Skills and culinary practices … differentiate one area from another. They are an integral part of individual, collective and territorial identity construction" (1998: 29). Women in rural and peri-urban areas of Oaxaca learn how to select the best utilitarian ceramic for their kitchen as part of their overall domestic education, and many will continue to buy from the same artisans as their mothers. Trips to the regional market are as much about these long-term relationships between vendors and customers as they are about the actual products purchased. Ceramic, like food ingredients, reflects the local tastes that are drawn from the environment.

Utilitarian ceramic cooking pots are a cottage craft and are made from local clays, formed through a traditional coil method without a potter's wheel, and fired in adobe kilns. Every ceramic producing community has what they call a *mina*, or area where they mine for clay, that they will transform into the particular style of pottery that is their area of expertise. Each family has a particular pot that forms their *oficio* or trade such as *comales, cazuelas, ollas, urnas, vasijas*, as well as ornamental ware. Unglazed *comales* (large clay plates with a slight curve) are used over open flames to prepare tortillas, adding the taste of earth and smoke to each tortilla. Cazuelas and ollas, as well as bowls, plates, and cups, are glazed after the first firing and returned to the adobe kiln for a second firing in order to create a glass-like finish that prevents liquids from seeping through them. Fire, water, clay, and a unique mixture of ingredients that form the glaze, come together in the kitchen of the Oaxacan cocinera to form the taste of Oaxaca.

The use of these beautiful ceramic vessels to prepare dishes that require extended time to cook such as beans, rice, broth, moles, stewed meat dishes, and other such daily dishes, provides tastes that cannot be replicated through metal or glass. Walk into most restaurants in Oaxaca and you will find these picturesque pieces on the burners, setting the stage for your meal as both a gastronomic event for your touristic experience (Mak, Lumbers and Evans 2011) and as verification that the dishes derive from an "authentic" heritage (Bessière 1998). Ceramic cooking utensils are part of the long tradition of food preparation that melds the old with the new and are instrumental to the unique taste of Oaxaca; their significance cannot be understated and defies replacement with modern tools, even if they bring great harm. These striking instruments of glazed, utilitarian ceramic are the primary culprit in the extensive lead poisoning that is found throughout the general population of Oaxaca and among Oaxaqueños in the United States (Azcona-Cruz et al. 2000; Grieshop 2006; Handley and Grieshop 2007; Handley et al. 2007; Hernández-Serrato et al. 2003; Pérez 2007; Pérez, Handley and Grieshop 2010; Preston 1996; Villalobos et al. 2009).

Traditional practices as identity

The issue of lead-based glazes in Oaxacan ceramic production is not new and has been a topic for researchers, journalists, and tourists for decades (Azcona-Cruz et al. 2000; Godines and McKim 2004; Handley and Grieshop 2007; Handley et al. 2007; Hernández-Serrato et al. 2003; Pérez 2007; Pérez, Handley and Grieshop 2010; Preston 1996; Villalobos et al. 2009). In fact, the pervasiveness of lead-based utilitarian ceramic and its incorporation into long-term food practices is evidenced by the issue of lead poisoning being explored as a health malady as early as the 1890s by Dr. Ruiz-Sandoval (Azcona-Cruz et al. 2000). While few tourists and others who ingest small quantities of lead through interim exposure, such as through the consumption of dishes while on vacation in Oaxaca, will rarely have a negative reaction to lead contamination, the ceramic artisans and their clients have demonstrated severe implications from lead poisoning. These maladies range from intestinal distress, anemia, erratic behavior, deafness, and severe learning disabilities among children to kidney failure, pancreatitis, high maternal mortality, early onset dementia, trends toward violent behavior, and infertility, among many other such issues, for adolescents and adults (Bellinger and Rappaport 2002; Hernández-Serrato et al. 2003; Needleman et al. 1990; Pérez 2007; Pérez, Handley and Grieshop 2010). These outcomes mirror many of the maladies associated with malnutrition and poverty, two issues that have defined the histories of most Oaxaqueños (Pérez 2007; Pérez, Handley and Grieshop 2010). As a result, many Oaxaqueños will deny poor health correlations to lead or will deny that lead is still a part of ceramic production (Pérez 2007; Pérez, Handley and Grieshop 2010) in order to justify the continued production and use of lead-based glazed ceramic. For ceramic producers, a shift away from lead oxide as the primary fluxing agent in their glazes is not economically viable as the low-lead glazes available to them are far more expensive and require higher temperatures and longer firing times in the kiln, raising the cost of ceramic production beyond the marketable price for their ware. In addition, the copper base in the low-lead glazes causes many foods to spoil quicker, produces a less-durable finish, and offers only a slight improvement in taste over glass or metal pots and pans. Ultimately, in my interviews with Oaxacan women, the continued use of lead-based ceramic is both an economic issue and a significant element of their identity that is linked through the traditional practices of the kitchen that in turn produces the sweet, smoky taste of Oaxacan cuisine.

The sweet, smoky taste of Oaxaca

For many of us, the ability to name particular tastes that form the aromatic structure of a dish is complicated. We are not professional tasters, who are primarily food scientists that undergo extensive training in culinary schools or obtain degrees in chemistry, nutrition, or other similar disciplines. Yet, many of us develop a fairly astute palate that can extract basic flavors from a dish. For me, the reality of this distinction came through a nutritional outreach project that I initiated in order to combat the absorption of lead (Pérez, Handley and Grieshop 2010). No matter how hard my team of local cocineras and I tried, we could not mimic the "taste" that was missing when we prepared local dishes with nonceramic instruments (Pérez 2007). The dozens of local men and women who formed our official "tasters," while enjoying the free food and opportunity to gather with friends in the afternoons, were kind in offering critiques that the foods were delicious but not authentic. Their comments included, "Se parece igual pero falta el sabor de acá" (It looks the same but it doesn't taste like food from here), "Tiene un buen sabor pero es más como allá" (It tastes good but it's more like [a dish] from there [referring to the United States]), and "Sale bien para los turistas pero prefiero el caldo de mi esposa" (It turned out well but I prefer the taste of my wife's soup). After weeks of struggling with these comments, the definition of this missing taste was finally captured by the head cocinera in our group, who described it as both "dulce como una pasa y ahumado como un chile colgado sobre el fuego del comal" (sweet like a raisin and smoky like a chili hung over the fire of the comal); her insight received an ovation and yells of "sí, eso!" (yes, that's it!) and "exactamente!" (Exactly!) among the other fourteen cocineras of our group.

To confirm our description of the taste, I set out to try and garner agreement that this is what differentiates food prepared in ceramic with that of other utensils. In addition to the local community where we were experimenting with recipes, I initiated informal interviews with other local Oaxaqueños as well as self-proclaimed aficionados of Oaxacan cuisine, including displaced Oaxaqueños in Baja California and California. I asked them to describe the difference in dishes that were prepared outside of Oaxaca in nonceramic cooking pots (Pérez 2015). My questions usually followed after we had consumed a traditional dish of Oaxaca or in reference to a known and appreciated dish and were basic: "Does this dish taste like the dish when you ate it in Oaxaca?" and "If not, why doesn't it taste the same as the dish you ate in Oaxaca?" For most tourists of Oaxaca, the answer was not a flavor they could name but rather was described as an essence that they felt lay in the atmosphere of being on vacation and in a special place. For displaced Oaxaqueños, it was taste but it was tied up in a sentimental reaction relative to loss and longing of a lifestyle that could not be replicated in their new communities. I found the explanation from an elderly gentleman from the Mixteca Baja region very poignant. He noted that at first he thought the problem was that his wife's heart was so heavy with sadness from living in the United States that she had forgotten how to cook, but later he decided that it was not her but rather the change of place, "Llevé las semillas de mi pueblo, pero el suelo y la lluvia no olían igual y el sol no besaba las plantas de la misma manera, entonces, la fruta no era la misma" (I carried the seeds from my community but the soil and the rain did not smell the same and the sun did not kiss the plants the same, so, the fruit was not the same [R. L. Pérez, unpublished fieldnotes, March 2014]). I felt frustrated that I could not get more distinct descriptions of taste, even though I realized that food and nostalgia are so intimately linked that I was asking them to do something unnatural—to separate the gusto of the cuisine that marked their memories from the process of tasting a dish. Sutton cites Kapella in calling out this kind of desire for the taste of home that cannot be found or replicated "as a 'burning of the lips' that comes from missing something deeply" (Kapella cited in Sutton 2001: 79). Sutton goes on to note that in his fieldsite of Kalymnos in Greece the noun *kaïmo* is used to describe the desire of displaced Kalymnians to have the taste of home through food, because it translates as both "psychic pain" and "uncontrollable desire" (Sutton 2001: 79). Although I could not draw out the description of difference as sweet and smoky, once I offered the description I tended to draw nods and agreement.

The Oaxacan migrant women living in San Quintín in Baja California were a bit more willing to disassociate longing from their memories. Like the elderly Oaxacan man in California, they also knew that the difference in taste was tied to the ingredients produced locally, including the taste produced from cooking with ceramic (Pérez 2015). They shared with me that when they were preparing food for weddings or other major celebrations, they would plan ahead of time to order the ceramic cooking vessels, along with the dried chili peppers and other important ingredients, from their families in Oaxaca who would have them sent on an *ADO* bus (*Autobuses de Oriente*, East line buses) that traveled between the two places. The demand for these instruments and ingredients has resulted in a whole new small business of cargo trucks that carry supplies from Oaxaca to San Quintín. For displaced Oaxaqueños in Baja California and the United States, the quest for home through the replication of taste is ultimately about identity and reinventing community through food in their new locations (Horowitz 2014; Maheshvari and Nzuza 2013; Mares 2012; Parasecoli 2014; Sutton 2001).

The movement of foods such as dried chili peppers, chapulines, *pasta de mole*, *salsa de chicatanas*, *tlayudas*, and other such staples to the Oaxacan palate from home communities in Oaxaca to the United States that I described earlier, resulted in an epidemic of lead poisoning in 2007 (Handley and Grieshop 2007; Handley et al. 2007). While low levels of lead ingestion may produce only minor physical symptoms, such as stomach cramping and diarrhea, the splurging on these foods in large quantities over a short time period that is similar to grand feasts resulted in severe enough symptoms to trigger concern at the local health clinic. Interviews with the families followed by blood lead-level testing confirmed sudden and intense exposure. More intensive research by the clinic resulted in the discovery of high lead levels among many residents, especially children, which was correlated to the foods that they were receiving from family members (Handley and Grieshop 2007).

Colleagues from northern California joined me in Oaxaca and together we spent several years creating binational and local education outreach programming on the hazards of lead poisoning along with nutrition intervention programs, and a focused intervention with local ceramic artisans on testing alternative glazes and firing techniques in order to eliminate the use of lead as a fluxing agent in their ceramic (Handley and Grieshop 2007; Pérez 2007; Pérez, Handley and Grieshop 2010). While all of these interventions had an immediate impact as people wrestled with the realizations of lead poisoning, ultimately, we could not replicate the tastes that are drawn from ceramic and that define the taste of Oaxaca for so many people. Local Oaxacan communities in northern California have accepted the new tastes that derive from local ingredients as a daily norm, and they are cautious with the gifts of food from their families in Oaxaca. In

Oaxaca, however, within a few short years, lead-based ceramic crept back into the kitchens of most families, maintaining the tastes of the past as the tastes of the present and, most likely, future.

In many ways, such adherence to all elements of cuisine—the use of ancient foodstuffs as the foundation of taste, traditional practices of farming and food preparation, and the centrality of it to identity in Mexico—is exactly what UNESCO captured in the honor. While anthropologists have known for some time that food is an exceptionally strong element of identity, we are learning the degree to which this may not be as malleable as other aspects of identity (Cantarero et al. 2013; Fischler 1988; Johnson 2016; Omori 2017; Pilcher 1998; Stead et al. 2011; Tibère 2016). That is, the gusto derived from food that satiates the body (outer self) and the mind (inner self) when it is inextricably linked to a definition of who we are, has such a strong hold over the way in which we understand ourselves that even the threat of death cannot separate us from the gusto we obtain through food.

Of equal importance to the recognition of food as identity is the challenge that such practices have on the future of Oaxaqueños. The Mexican state has known for generations that the ceramic produced by their rural artisans is laden with this silent and debilitating toxin (Azcona-Cruz et al. 2000; Godines and McKim 2004; Handley and Grieshop 2007; Handley et al. 2007; Hernández-Serrato et al. 2003; Pérez 2007; Pérez, Handley and Grieshop 2010; Preston 1996; Villalobos et al. 2009) but also has chosen not to address it as a significant health malady (Pérez 2007). The state also knows that tourism has provided these same communities with economic stability, and as I have argued elsewhere, the female producers are willing to accept the health consequences from lead in order to retain the economic and political autonomy that their craft has brought them (Pérez 2007). The recognition of the cuisine of Mexico by UNESCO reinvigorated tourism to Oaxaca and, in the process, reified not just the chefs and restaurants in the valley and coastal tourist centers, but also the marginal producers and farmers who have kept these ancient foodstuffs alive along with the women who produce and use traditional instruments of the kitchen, both of which define the taste of Oaxaca.

Note

1 Oaxaca is the name of both the state in the federal republic, and the capital city of the same state.

References

Azcona-Cruz, M. I., S. J. Rothenberg, L. Schnaas, J. S. Zamora-Muñoz and M. Romero-Placeres (2000), "Lead-glazed Ceramic Ware and Blood Lead Levels of Children in the City of Oaxaca, Mexico," *Archives of Environmental Health*, 55 (3): 217–222.

Baldacchino, G. (2015), "Feeding the Rural Tourism Strategy? Food and Notions of Place and Identity," *Scandinavian Journal of Hospitality and Tourism*, 15 (1/2): 1–16.

Bellinger, D. and L. Rappaport (2002), "Developmental Assessment and Interventions," in B. Harvey, ed., *Managing Elevated Blood Lead Levels among Young Children: Recommendations from the Advisory Committee on Childhood Lead Poisoning Prevention*, 78–95, Atlanta, GA: US Department of Health and Human Services.

Bessière, J. (1998), "Local Development and Heritage: Traditional Food and Cuisine as Tourist Attractions in Rural Areas," *Sociologia Ruralis*, 38 (1): 21–34.

Bourdieu, P. ([1979] 1984), *Distinction: A Social Critique of the Judgment of Taste*, trans. R. Nice, Cambridge, MA: Harvard University Press.

Cantarero, L., E. Espeitx, M. Gil Lacrus and P. Martín (2013), "Human Food Preferences and Cultural Identity: The Case of Aragón, Spain," *International Journal of Psychology*, 48 (5): 881–890.

Cocking, L. (2017), "Is Oaxaca Mexico's New Culinary Hotspot?" *The Culture Trip*, https://theculturetrip.com/north-america/mexico/articles/is-oaxaca-mexicos-new-culinary-hotspot/ (accessed January 8, 2018).

Everett, S. and C. Aitchison (2008), "The Role of Food Tourism in Sustaining Regional Identity: A Case Study of Cornwall, South West England," *Journal of Sustainable Tourism*, 16 (2): 1–167.

Fischler, C. (1988), "Food, Self, Identity," *Anthropology of Food/Anthropologie de l'alimentation*, 27 (2): 275–292.

Gilbert, S. (2014), "Oaxaca: The Gourmet Heart of Mexico," *Lonely Planet*. www.lonelyplanet.com/travel-tips-and-articles/oaxaca-the-gourmet-heart-of-mexico/40625c8c-8a11-5710-a052-1479d276138f (accessed January 8, 2018).

Godines, V. and J. B. McKim (2004), "Part 4: Poisoned Packages: Register Finds Lead-Poisoned Children in Mexican Village Where Pots are Made for Candy," *Orange County Register*, April 28, 2004, www.ocregister.com/articles/toxic-219227-laguna-treats.html (accessed September 27, 2010).

Grieshop, J. I. (2006), "The Envíos of San Pablo Huixtepec, Oaxaca: Food, Home and Transnationalism," *Human Organization*, 65 (4): 400–406.

Handley, M. A. and J. I. Grieshop (2007), "Globalized Migration and Transnational Epidemiology," *International Journal of Epidemiology*, 36 (6): 1205–1206.

Handley, M. A., C. Hall, E. Sanford, E. Diaz, E. Gonzalez-Mendez, K. Drace, R. Wilson, M. Villalobos and M. Chroughan (2007), "Globalization, Binational Communities, and Imported Food Risks: Results of an Outbreak Investigation of Lead Poisoning in Monterey County, California," *American Journal of Public Health*, 97 (5): 900–906.

Hernández-Serrato, M. I., L. R. Mendoza-Alvarado, R. Rojas-Martinez, C. González-Garza, J. M. Hulme and G. Olaiz-Fernandez (2003), "Factor Associated with Lead Exposure in Oaxaca, Mexico," *Journal of Exposure Analysis and Environmental Epidemiology*, 13 (5): 341–347.

Holtzman, J. D. (2006), "Food and Memory," *Annual Review of Anthropology*, 35: 361–378.

Horowitz E. (2014), "Remembering the Fish and Making a Tsimmes: Jewish Food, Jewish Identity, and Jewish Memory," *The Jewish Quarterly Review*, 104 (1): 57–79.

Johnson, M. C. (2016), "'Nothing is Sweet in my Mouth': Food, Identity and Religion in African Lisbon," *Food & Foodways*, 24 (3): 232–254.

Lin, Y.-C., T. E. Perason and L. A. Cai (2011), "Food as a Form of Destination Identity: A Tourism Destination Brand Perspective," *Tourism and Hospitality Research*, 11 (1): 30–48.

Lorence, D. H. and A. Garcìa Mendoza (1989), "Oaxaca, Mexico," in D. G. Campbell, ed., *The World Wildlife Fund World Strategy for the Inventory of Tropical Forests*, New York: New York Botanical Garden.

Maheshvari, N. and N. Nzuza (2013), "Food and Maintaining Identity for Migrants: Sierra Leone Migrants in Durban," *Journal of Sociology and Social Anthropology*, 4 (3): 193–200.

Mak, A. H. N., M. Lumbers and A. Evans (2011), "Globalisation and Food Consumption in Tourism," *Annals of Tourism Research*, 39 (1): 171–196.

Mares, T. M. (2012), "Tracing Immigrant Identity through the Plate and Palate," *Latino Studies*, 10 (3): 334–354.

Needleman, H. L., A. Schell, D. C. Bellinger, A. Levinton and E. N. Allred (1990), "The Long-term Effects of Exposure to Low Doses of Lead in Childhood: An 11 Year Follow-up Report," *New England Journal of Medicine*, 322 (2): 83–88.

Omori, H. (2017), "Eating Japanese Food in Diaspora as Identity Building: The Case of a Japanese Canadian Church," *Contemporary Japan*, 29 (2): 148–161.

Parasecoli, F. (2014), "Food, Identity, and Cultural Reproduction in Immigrant Communities," *Social Research*, 81 (2): 415–440.

Pérez, R. L. (2007), "Challenges to Motherhood: The Moral Economy of Oaxacan Ceramic Production and the Politics of Reproduction," *Journal of Anthropological Research*, 63 (3): 305–330.

Pérez, R. L. (2015), "If you don't use chiles from Oaxaca is it still Mole Negro? Shifts in Traditional Practices, Techniques and Ingredients of Cuisine among Oaxacan Migrants," in S. I. Ayora-Diaz, ed., *Cooking Technology: Transformations in Culinary Practice in Mexico and Latin America*, 99–110, London: Bloomsbury Academic.

Pérez, R. L., M. A. Handley and J. Grieshop (2010), "Savoring the Taste of Home: The Pervasiveness of Lead Poisoning from Ceramic and its Implications in Transnational Care Packages," *NAPA Bulletin*, 34 (1): 105–125.

Pilcher, J. M. (1998), *Que Vivan Los Tamales! Food and the Making of Mexican Identity*, Albuquerque: University of New Mexico Press.

Preston, J. (1996), "Santa Maria Atzompa Journal: Pots that Poison and Potters Facing Broken Lives," *New York Times*, International Section, June 14, 1996.

Rubenstein, G. (2012), "The Pop-up Food Shops of Oaxaca," *The Atlantic*, www.theatlantic.com/health/archive/2012/04/the-pop-up-food-shops-of-oaxaca/255347/ (accessed January 8, 2018).

Stead, M., L. Mcdermott, A. M. Mackintosh and A. Adamson (2011), "Why Health Eating Is Bad for Young People's Health: Identity, Belonging and Food," *Social Science & Medicine*, 72 (7): 1131–1139.

Sutton, D. (2001), *In Remembrance of Repasts: An Anthropology of Food and Memory*, Oxford: Berg.

Tibère, L. (2016), "Food as a Factor of Collective Identity: The Case of Creolization," *French Cultural Studies*, 27 (1): 85–95.

UNESCO (United Nations Education, Scientific and Culture Organization) (2010), Convention for the Safeguarding of the Intangible Cultural Heritage. Intergovernmental Committee for the Safeguarding of the Intangible Cultural Heritage, Fifth Session, Nairobi, Kenya, November 15–20, 2010, https://ich.unesco.org/doc/src/ITH-10-5.COM-CONF.202-6-EN.pdf (accessed August 12, 2017).

Villalobos, M., C. Merino-Sánchez, C. Hall, J. Grieshop, M. E. Gutierrez-Ruiz and M. A. Handley (2009), "Lead (II) Detection and Contamination Routes in Environmental Sources, Cookware, and Home-prepared Foods from Zimatlán, Oaxaca, Mexico," *Science of the Total Environment*, 407 (8): 2836–2844.

Von Bremzen, A. (2016), "More than Mole: Why Oaxaca is Mexico's Hottest New Food Destination," *Travel and Leisure*, www.travelandleisure.com/articles/oaxaca-mexico-newest-food-town (accessed January 8, 2018).

10

Dos Equis and Five Rabbit: Beer and Taste in Greater Mexico

Jeffrey M. Pilcher
University of Toronto

Dos Equis beer, created by the Moctezuma Brewery in 1897, embodied the Mexican ideal of industrial modernity. Originally referred to as Siglo XX (Twentieth Century), the brand heralded a progressive future of European-style consumption while, nevertheless, maintaining patriotic ties to the past with an image of the Aztec emperor on the label, staring out from between the two "X"s. Mexican leaders hoped that pasteurized bottles of European beer would displace the smelly, viscous, native brew, *pulque*, which was personified by indigenous deities of drunkenness such as Macuiltochtli (Five Rabbit). Over the course of the twentieth century, Dos Equis and other lager beers indeed became the national drink, an inevitable accompaniment to popular celebrations. And as the masses consumed ever-greater quantities of light, fizzy beer, the logic of social distinction inspired trendy millennial Mexicans to revive the strongly flavored pulque, even as the first Latino craft brewery in the United States was named 5 Rabbit.

The sociologist Pierre Bourdieu ([1979] 1984) has explained how education and culture have bolstered distinction and privilege in the face of economic change and social mobility. In Mexico City, as in Paris, knowledge of French haute cuisine represented a form of cultural capital that helped aristocrats maintain wealth and status through a century of revolutionary upheaval. By the late twentieth century, France had lost its cultural hegemony through the fragmentation of global standards of taste, and a new "omnivorous" elite emerged that was equally comfortable with the French gastronomical meal and Mexican regional cuisine, thereby synthesizing bourgeois taste with its bohemian antithesis. But as the sociologists Josée Johnston and Shyon Baumann (2010) observed, the omnivore did not so much democratize taste as raise the bar of cultural knowledge from a single, albeit expensive, restaurant culture to familiarity with and leisure to travel among all the world's cuisines. Indeed, newly desirable foods such as pulque, with its off-putting smell and texture, exemplified the elite refusal of accessible pleasures that Bourdieu identified.

Nevertheless, the sociological analysis of taste as social distinction must be complemented by taking into account a historical perspective on the sensory experience of taste. Critics have challenged the static nature of Bourdieu's analysis and his assumption that sensory experience is a function of social hierarchy (Hennion 2007). While food science research has confirmed that social cues can influence the perception of taste (Allison and Uhl 1964), this follows from rather than determines the complex interplay between senses and memory that constitutes the experience of flavor (Shepherd 2011). The anthropologist Steffan Igor Ayora-Díaz (2012) has argued that regional cuisines emerge from a historical process of "naturalization of taste" through the repetition and exchange of recipes, both local and cosmopolitan. Drinking cultures, like regional cuisines, can also be significant zones of cultural encounter and taste formation (Wilson 2005).

In the case of Mexican beer, the historians Susan Gauss and Edward Beatty (2014) have noted three crucial turning points: the rise of a local industry based on foreign technology around 1890, the national consolidation of markets and expertise beginning in the 1920s, and the global expansion of Mexican exports from the 1980s. This ongoing interplay between local and global brewing indicates the need for a transnational analysis of what the Texas folklorist Américo Paredes (1976) called "Greater Mexico," a region of Mexican cultural influence that now spans large parts of North America. Communities of taste for Mexican beer took shape historically through sensorial encounters and social distinctions that crossed national and ethnic boundaries.

Pulque: From tradition to modernity

The inhabitants of Mesoamerica, like most people around the world, embraced the sociability of drinking together while fearing the social disorder that came from drinking too much (Heath 2000). Pulque, fermented from the sap of the maguey plant, was associated with both the goddess of fertility and nourishment, Mayahuel, and the gods of drunkenness and disorder, Centzon Totochtin (Four Hundred Rabbits), among which Five Rabbit was particularly feared. Successive rulers sought to restrict the consumption of pulque, from Aztec emperors to the viceroys of New Spain and the governors of the Republic of Mexico. Yet they simultaneously benefited from it, through social distinction before the conquest and as a lucrative source of revenue afterwards, and as a result the beverage remained fundamental to popular culture and diet.

Archaeological evidence of pulque production dates to the great preclassical city of Teotihuacán (*c.* 100 BCE to 700 CE), and the drink had spread widely throughout the *altiplano* (highland) by the postclassical era (Correa-Ascencio et al. 2014). Once the agave plant had reached maturity, after about seven years, a skilled tapper (*tlachiquero*) harvested the sap (*aguamiel*) twice daily. The pulque was usually consumed after a few days of fermentation, although roots, herbs, and bark were often added as preservatives and flavoring. Consumption seems to have taken place in the context of communal religious celebrations, but our knowledge of native drinking cultures is colored by the puritanical Mexica, a nomadic tribe who founded the Aztec Empire and harshly restricted pulque for all but the elite and the elderly, except on ritual occasions. Despite the evidence of codices written by the Hispanified offspring of Mexica nobles, the ease of making pulque and its dietary importance suggest that the beverage was drunk regularly, if covertly, outside the capital Tenochtitlan (Guerrero Guerrero 1980).

Aztec fears of drunkenness persisted under Spanish colonialism, as pulque became associated with native delinquency. The historian William Taylor (1979) concluded that although pulque consumption did rise after the conquest, the Spaniards' primary objection to indigenous drinking was its ecstatic, religious nature, in contrast to the European ideal of self-control. In any event, the Crown made only desultory attempts to restrict pulque because Spanish aristocrats owned the leading agave plantations and the colonial government depended on alcohol taxes. While Spanish elites dominated the commerce, native women did most of the actual marketing, both within rural communities and in bringing pulque from the countryside to Mexico City. By the late eighteenth century, the capital's fringes bustled with taverns (*pulquerías*), and American-born Spaniards (*criollos*) shared in this community of taste. The scientist Alexander von Humboldt (1836: 338) compared it to cider with "a smell of rotting meat," but he noted that criollos "who have conquered the disgust caused by the fetid odor prefer pulque to all other beverages."

This ambivalent attitude continued through the first century of Mexican independence; the historian Áurea Toxqui has shown that even as elites came to view drunkenness as the primary cause of national backwardness, pulque became an expression of working-class modernity. Taverns were an important space for popular political mobilizing during the freewheeling elections of the early republic, and liberal intellectuals often joined in these discussions. Manuel Payno (1864: 438) listed the types available—"fine pulque, sweet pulque, strong pulque, ordinary pulque, *tlachique*"—recommending that consumers seek out *fino*. Manuel Orozco y Berra (1855: 361) described the lowest grade, *tlachique*, as "bad tasting, slimy (baboso), and thick." As midcentury governments limited the franchise and sought to clean up the city center, pulquerías colonized the periphery, introducing urban patterns of sociability to formerly rural communities. With the rise of industrialization under the government of Porfirio Díaz (1876–1910), Mexico City began attracting large numbers of rural migrants, who sought out the familiar spaces and tastes of pulquerías as a respite from long hours working in factories and peddling. But while pulque became part of the urban experience for newcomers, the Porfirian elite considered migrants and their beverage as a foul-smelling blight on the modern city (Toxqui 2011, 2014).

Dos Equis: The taste of modernity

The historiography of beer in Mexico emphasizes the social distinction of beer's European origins. Authors frequently begin their narratives with Mexico's legendary first brewery, founded by Alonso de Herrera in 1544 following fashions set in the court of Charles V, who had grown up drinking beer in his native Ghent before ascending the Spanish throne (Gauss and Beatty 2014). The Austrian Archduke Maximilian, installed briefly as emperor by French invaders in the 1860s, is also often cited as a patron of Mexican brewing (Ruiz 2011). But despite its European roots, Porfirian brewing became a thoroughly modern industry conducted along the lines of giant firms in the United States.

The first beers available in nineteenth-century Mexico were dark Dublin stouts and India Pale Ales (IPAs), which the British marketed globally and not just to their colonists in India. These top-fermented ales were as strongly flavored in their own way as Mexican pulque—the stouts dominated by bittersweet coffee and caramel malts and IPAs with bitter hop notes among wild yeast flavors of *Brettanomyces*, often described as "barnyard" or "mousy" (Bamforth 2009: 65). By contrast, beginning in the 1890s Mexican industrial brewers preferred Bavaria's unique bottom-fermenting lager yeasts, which worked at cooler temperatures and therefore posed less risk of spoilage from bacterial contamination, especially after the introduction of pure yeast strains. Lagers ranged across three principal styles, depending on the malt: full-bodied, dark Munich; mellow, amber Vienna; and dry, golden Pilsner.

The regionalism of early Mexican brewing resulted more from the limitations of urban markets and poor transportation than from distinctive local tastes, which only emerged over time. The historian Gabriela Recio (2007) found that although some thirty breweries operated at the turn of the century, 70 percent of national production came from just five firms located in leading industrial centers. In addition to Moctezuma in the petroleum and textile-producing state of Veracruz, these firms were Cuauhtémoc, in the steel town of Monterrey; Toluca y México, near the nation's capital; and Chihuahua and Sonora, northern mining states. Each brewery united local capital and foreign technicians: at Cuauhtémoc, the Garza Sada clan partnered with brewer Joseph Schnaider, while Moctezuma was associated with Finance Minister José Yves Limantour and brewer Philippe Suberbie.

Newspaper advertisements, especially in the glossy *El Mundo Ilustrado* (The World Illustrated), document the aspirations of Porfirian drinking cultures for European sophistication, modern sanitation, and social uplift. Brewers made quite specific references to Munich and Pilsner styles; Cuauhtémoc's Salvator represented the former while its Bohemia brand followed the latter. Toluca's Victoria claimed to be the preferred beer of the German colony of Mexico City. But despite the elite's nostalgic memories of Maximilian's ill-fated empire, there are no known references to the Viennese brewing style. Advertisements described XX as simply lager, although differentiated from the dark XXX brand introduced by Moctezuma in 1906. Only highly skilled workers could afford beer on a regular basis, but Mexican brewers also offered more affordable products to inculcate a popular taste for beer. Cuauhtémoc's Estrella brand was described as the "favorite of the popular classes" ("La Cervecería Cuauhtemoc" 1908: 845). Newspapers also emphasized the strict hygiene of industrial beer production, with images of modern factories in contrast to the uncontrolled fermentation of pulque. Consumers, for their part, cultivated a sense of cosmopolitanism around beer drinking. In Mérida, where a chilled lager offered a welcome break from the midday heat, polite society adopted the "lunch cerveza" ("Carnet Social" 1918).

Despite allusions to Europe, Mexican market conditions had little in common with the low wages and demanding consumers of nineteenth-century Germany. Instead, brewers followed the example set in St. Louis and Milwaukee of laborsaving machinery and barley malt adjuncts such as maize to speed production of light lagers. Technicians generally had training and experience in the United States; Cuauhtémoc's Joseph Schnaider had formerly operated a St. Louis brewery while Luis G. Sada studied at the Wahl and Henius Brewing Academy of Chicago, as did Moctezuma's Philippe Suberbie. Likewise, the preferred variety of brewer's yeast in Mexico was Wahl and Henius's pure lager culture, known as Chicago Number 1. A school brochure proudly declared: "The brewing industry of Mexico, which had been conducted strictly according to German models, was Americanized, adopting infusion methods, and at the present time nearly all Mexican breweries are operated by American brewmasters" (American Brewing Academy of Chicago 1901: 146; see also Womack 2012). North American influence likely also explains the Mexican preference for amber beers, which was intended not to reproduce Viennese-style lagers but as a cheaper method of brewing dark beers without the expense of full-bodied Munich malts. To this day, most Mexicans classify beer according to color, as either light or dark, and that in itself marks another significant change from the exclusivity of Porfirian beer culture.

Corona: From Mexico to the world

In 1943, the Cervecería Modelo (Model Brewery) began a long-lived advertising campaign for its Corona brand with the slogan: "Twenty million Mexicans can't be mistaken!" (Herrero 2001: 45). The claim that the entire national population preferred Corona, however presumptuous, differed from the elitist associations of beer during the Porfiriato. Mass marketing made sense in the decades around midcentury, as postrevolutionary economic prosperity and urban growth made beer affordable for factory workers and other middling sectors. Import substitution policies and improved transportation and communication also favored the consolidation of regional Porfirian breweries. National markets in turn provided a basis for global expansion beginning in the 1980s, when economic and social crises encouraged firms to look abroad for future growth. The centrality of Mexican identity in 1940s advertisements had disappeared completely by 1992, when Modelo rolled out its "change your whole latitude" campaign depicting Corona as an anonymous beach vacation in a bottle.

Rivalry between Monterrey's industry-leading Cuauhtémoc and the Mexico City upstart Modelo drove the midcentury nationalization of Mexican brewing. Founded in 1925 by wealthy Basque bakers, Modelo introduced a range of brands, including the short-lived Moravia, intended to compete with Cuauhtémoc's renowned Bohemia. But Old World charm had limited appeal for postrevolutionary consumers, and Modelo built a larger, more proletarian market for its Corona brand through a combination of modernist sensibility (art deco lettering) and nationalist symbolism (the combination of a Spanish crown and an indigenous feathered serpent), painted on clear, glass bottles. When the fledgling Modelo rebuffed a buyout offer, Cuauhtémoc instead purchased a Mexico City rival, the Cervecería Central, in 1928. A decade later, the Monterrey-based giant had consolidated a national presence through mergers and contract brewing. Modelo then embarked on its own expansionist plan, acquiring the Cervecería de Toluca y México in 1935, taking over the Estrella and Pacífico breweries in the 1950s, and building new plants in Sonora and Coahuila in the 1960s. By the 1980s, Cuautémoc's merger with Moctezuma and Modelo's purchase of the Cervecería Yucateca cemented a duopoly within the Mexican beer industry (Gauss and Beatty 2014).

Nationalization also entailed the training of brewers and sourcing of raw materials, thereby breaking the dependence on North American and European expertise and supply chains. Brewing technology was one of the original courses of study at the Instituto Tecnológico de Monterrey, which was modeled on the Massachusetts Institute of Technology in 1943 by Cuauhtémoc's owners, the Garza Sada family. In the 1960s, the company's head of malting, Guillermo Ceballos Aguilera, devised new machinery to process and malt sorghum, which had become a major fodder crop in the neighboring state of Tamaulipas as part of the "Green Revolution" of agricultural modernization. Thereafter, selling cheap domestic malt to rival Mexican breweries became a valuable source of revenue for Cuauhtémoc (Hibino 1992).

Nationalization also had significant consequences for the taste of beer through mass marketing. As a social construct, the taste of drinking occasions became more commercialized; for example, the Yucatecan "cerveza lunch" was transformed into the "Crystal Hour" (*Hora Cristal*), named after a popular radio program (Ayora-Díaz and Vargas Cetina 2005). Advertisers appealed to broad audiences through claims of "twenty million Mexicans" as well as by employing celebrity spokespersons with a common touch such as Mario Moreno "Cantinflas," Pedro Infante, and Celia Cruz. As a sensory experience, the taste of beer became generally lighter on a global scale at midcentury, and Mexican versions had particular national characteristics. The widespread use of sorghum malts for working-class brands pioneered a technology that was later adopted worldwide for gluten-free beers, although many consumers found them distasteful. Likewise, Corona's clear glass bottles increased the risk of light exposure, which can cause "degradation of iso-α-acid bitter substances," a condition "referred to either as 'lightstruck,' or worse still, 'skunky'" (Bamforth 2009: 87). National consolidation also led to the homogenization of regional styles dating to the Porfirian era. After Modelo took over the Cervecería

Yucateca, many customers complained that local brands such as Negra León and Montejo did not taste the same (Ayora-Díaz and Vargas Cetina 2005).

Corona first gained an international market among North American university students precisely because of its downscale associations in Mexico. Modelo's import agent introduced the brand to Southwestern markets in the early 1980s without particularly high hopes, and when sales unexpectedly took off the company was unable to meet demand. Regional shortages and tales of hoarding served to heighten the brand's cachet, and when production ramped up by 1986, Corona sold twelve million cases, second only to Heineken. Rival distributors, dismayed at their declining market share, spread rumors that the Mexican beer was contaminated by urine, a libelous claim they were forced by court order to retract. Even Modelo's executives were unable to explain their own success, apart from vague references to quality. Some of the first consumers in the United States had recognized the distinctive, clear-glass bottles from visits to Mexican border-town bars or surfing trips to Baja California. The beer's "cool" reputation was enhanced by celebrity early adopters such as the singer Jimmy Buffett, of "Margaritaville" fame, who was quickly hired to promote the brand. Bartenders described the novel behavior of Corona drinkers, learned on trips to Mexico, of personalizing the act of consumption by squeezing a slice of lime down the longneck, placing a thumb over the top, and inverting the bottle, so that the lime suffused the beer. But the most unexpected explanation came from young, male patrons themselves. When invited to focus groups by the company, they repeatedly declared: "It looks like another liquid" (quoted in Pilcher 2012: 179). Far from worrying about contamination, hip consumers laughed at the scatological joke.

By the end of the twentieth century, Mexican brewers had exceeded the wildest dreams of their Porfirian predecessors, not only building a nation of beer drinkers but also conquering global markets. True, the Mexican elite had moved on to more exclusive drinks such as Scotch whisky, but they accounted for only a miniscule share of the market. On a technical level, Mexican brewmasters had gone from acquiring foreign skills and machinery to innovators in their own right. Nevertheless, surviving in the face of changing tastes and global competitors would prove far more difficult.

5 Rabbit: Brewing postmodernity

The dominance of modern, homogenized lager beers within national markets led to new configurations of global and local industries and tastes around the turn of the millennium. At the global level, a new wave of consolidation swept through the industry, led by the 2004 merger of Brazil's Ambev and Belgium's Interbrew and its 2008 hostile takeover of Anheuser-Busch, creating the world's largest brewing firm, A-B Inbev (Howard 2014). Even as mass markets became increasingly globalized, consumers began postmodern local revolts against the homogenized taste of generic lager beers. In the United Kingdom, a group of beer lovers formed the Campaign for Real Ale in 1971 to revitalize the distinctiveness of local pubs and breweries. Meanwhile in the United States, a microbrewing movement sought to reinvent diverse traditional beers such as IPAs, wheat beers, and even farmhouse *saisons* made with wild yeast. In Mexico, the revitalization of native brewing traditions and the localization of international styles took place simultaneously as young people formed pulque-brewing collaboratives or experimented adding *chile* or agave to IPAs and stouts.

Modelo's success exporting Corona was only the first stage of a global process that soon culminated in foreign takeovers of the national beer industry. Beginning in 1994, the North American Free Trade Agreement allowed brewers to import large quantities of prime malt, which helped to satisfy a more demanding and fickle clientele than Mexico's working classes. Soon, both Modelo and Cuauhtémoc-Moctezuma were forced to offer minority stakes in their companies to survive in fiercely competitive global markets, and by 2013, they had sold out completely to A-B Inbev and Heineken, respectively. Foreign managers continued the pattern, begun with Corona, of de-Mexicanizing the brands. Heineken advertised Dos Equis with the highly successful "Most Interesting Man in the World" campaign, featuring an Ernest Hemingway-like figure with such improbable lines as: "When he goes to Spain, he chases the bulls" (Schultz 2012). Nevertheless, the primary goal of the acquisition had been to increase the share of the company's premium Heineken brand in the Mexican market.

Meanwhile, the nationalist revival of pulque traditions occurred within the context of culinary tourism and urban gentrification. Pulque production had remained flat from the 1930s to the 1980s, maintaining a loyal but aging clientele of rural and urban workers, while gaining few new converts among the rapidly growing population. The political crises of the 1990s seemed to mark the end, as agave acreage dedicated to pulque (as opposed to tequila) hit new lows and most of Mexico City's 1,500 pulquerías closed their doors (Mitchell 2004; Rojas Rivas et al. 2016). Even as the folklorist and organic intellectual Armando Jiménez (2000) wrote nostalgically of plebeian drinking cultures, Mexico's peasant and street foods gained international tourist attention, culminating in 2010 with UNESCO's declaration of Mexican regional cuisine as an intangible cultural heritage of humanity (Pilcher 2012). Fueled by tourist demand for authenticity, pulquerías sprang up in the trendy Mexico City neighborhoods Coyoacán and Colonia Roma and even as far afield as New York City and Chicago. Bearded Mexican hipsters began experimenting with new ingredients such as coffee, mint, and oats in place of more traditional *curados*—pulques cured with coconut, mango, and pineapple. Seeking to guide Mexican and foreign tourists to more rustic and seemingly authentic producers, government officials worked with local promoters to organize pulque fairs and trails.

Mexico's craft beer movement likewise forged cross-border ties of technique, supply, and markets. Gustavo González, regarded as the father of Mexican *cerveza artesanal*, began homebrewing in 1995 with North American inspiration. He launched commercial operations in 2001 under the Cosaca label, making just three basic ales: *negra, roja,* and *güera* (dark, red, and blond). Other craft beer pioneers were migrants, such as Gilbert Nielsen, who founded Calavera in 2008 after winning a Danish home-brewing contest (Morales n.d.). Although brewers experimented with local flavorings, for example, adding chile and chocolate to create a Mexican imperial stout, their use of imported hops and malt priced craft beers out of the reach of working-class consumers. Modelo and Cuauhtémoc-Moctezuma's control over distribution chains also prevented small brewers from placing their products in retail outlets, at least until the duopoly was taken over by foreigners. Highly sensitive to bad publicity, the new managers agreed to distribute craft beers and even offered to purchase select microbreweries in order to profit from the growing market (Partlow 2015).

Mexican craft beer has also begun to circulate back across the border to the United States. Breweries in Baja found a ready market for their products in San Diego and Los Angeles, although export licenses were daunting for small firms (Bennett 2015). In Chicago, the 5 Rabbit Brewery, founded in 2011 by Andrés Araya and Isaak Showaki, claimed to be the country's first Latin American-inspired craft brewery. In addition to the eponymous 5 Rabbit beer, a golden ale, they offered 5 Lizard, a Belgian *witbeer* (wheat beer) with lime substituted for the traditional orange peel, and 5 Vulture, a Oaxacan version of dark ale made with ancho chile and *piloncillo* (rustic brown sugar). The brewery gained widespread notice in June 2015, following racist slurs against Mexican immigrants made by then-presidential candidate Donald Trump. Having supplied a blond ale as a house beer for the Trump Tower in Chicago, 5 Rabbit cancelled the contract and sold the remaining inventory with the label "Chinga-Tu-Pelo" (Fuck Your

Hair). They no longer produce the beer, but continue to sell T-shirts and other memorabilia, donating profits to immigrant rights groups (Araya n.d.).

In just a few decades around the turn of the millennium, the market conditions, social hierarchies, and taste profiles of Mexican brewing have been transformed dramatically. The beer industry, a pioneer of national development since the Porfirian era and a seeming champion of global expansion, has fallen into foreign hands. Meanwhile, the easy-drinking taste of lager has become passé, while strongly flavored pulques and craft beers have been revived. Finally, craft brewers have Mexicanized their product with chiles, spices, and tropical fruits, just as New Age pulqueros added exotic touches to the native brew. While it is impossible to guess the next big trend, cultural exchange will surely continue to drive the taste of Mexican beer.

Conclusion

Taste as a sensory experience has evolved through centuries of foreign contact, but hierarchies of taste as social status have persisted from Mexica warriors and Spanish conquistadors to central European brewers and most recently North American microbrewers. Beer became associated with the taste of modernity in the nineteenth century through a combination of both sensory stimuli and social cues. Midcentury imports of dark stout and pale ale were highly valued because of their rarity and association with Britain, but their strong flavors were as much of an acquired taste as the indigenous pulque. With the introduction of mild, clear lager, beer's opposition to pulque became sharply apparent in taste as well as in hygienic factory production. Then, a century later, as lager beer became accessible to all levels of society and North Americans came to associate local beers with slumming, the Mexican elite turned instead to more strongly flavored beverages, pulque and craft beer.

With its homogenized, factory production, beer seemed to allow little initiative for consumers to adapt the product to their own tastes. The distinctiveness of Mexican beer varieties, for example, the use of relatively pale rather than full-bodied malts for "dark" beers such as Negra Modelo, probably arose from producers' cost calculations rather than consumers' taste preferences. Indeed, the industry's mid-twentieth-century national consolidation, which simultaneously differentiated Mexican beers from international rivals while homogenizing regional differences, often came at the expense of local tastes. Nevertheless, the popular classes transformed and localized industrial beer through their drinking cultures, for example, the habit of dosing it with lime, an inevitable condiment on the Mexican table, and devising new concoctions such as the *michelada*, a beer mixed with lime, Worcestershire sauce, chile, and spices. Such popular innovations, rather than corporate branding, made Mexican beer a global phenomenon.

References

Allison, R. I. and K. P. Uhl (1964), "Influence of Beer Brand Identification on Taste Perception," *Journal of Marketing Research*, 1 (3): 36–39.

American Brewing Academy of Chicago (1901), *Tenth Anniversary Reunion: Alumni and Former Students*, Chicago: American Brewing Academy of Chicago.

Araya, A. (n.d.), "Chinga-Tu-Pelo," www.5rabbitbrewery.com/chinga-tu-pelo/ (accessed June 15, 2017).

Ayora-Díaz, S. I. (2012), *Foodscapes, Foodfields, and Identities in Yucatán*, Amsterdam: CEDLA, Oxford: Berghahn.

Ayora-Díaz, S. I. and G. Vargas-Cetina (2005), "Romantic Moods: Food, Beer, Music, and the Yucatecan Soul," in T. M. Wilson, ed., *Drinking Cultures: Alcohol and Identity*, 155–178, Oxford: Berg.

Bamforth, C. (2009), *Beer: Tap into the Art and Science of Brewing*, 3rd edn., Oxford: Oxford University Press.

Bennett, S. (2015), "Where to Find Mexican Craft Beer in L.A.," *LA Weekly*, August 14, 2015, www.laweekly.com/restaurants/where-to-find-mexican-craft-beer-in-la-5673598 (accessed February 1, 2017).

Bourdieu, P. ([1979] 1984), *Distinction: A Social Critique of the Judgment of Taste*, trans. R. Nice, Cambridge, MA: Harvard University Press.

"Carnet Social" (1918), *Revista de Yucatán*, October 14, 1918: 4.

Correa-Ascencio, M., I. G. Robertson, O. Cabrera-Cortés, R. Cabrera-Castro and R. P. Evershed (2014), "Pulque Production from Fermented Agave Sap as a Dietary Supplement in Prehispanic Mesoamerica," *Proceedings of the National Academy of Sciences*, 111 (39): 14223–14228. https://doi.org/10.1073/pnas.1408339111.

Gauss, S. M. and E. Beatty (2014), "The World's Beer: The Historical Geography of Brewing in Mexico," in M. Patterson and N. H. Pullen, eds., *The Geography of Beer: Regions, Environment, and Societies*, 57–65, Dordrecht: Springer.

Guerrero Guerrero, R. (1980), *El pulque: Religión, cultura, folklore*, Mexico City: INAH.

Heath, D. B. (2000), *Drinking Occasions: Comparative Perspectives on Alcohol and Culture*, Philadelphia: Brunner/Mazel.

Hennion, A. (2007), "Those Things That Hold Us Together: Taste and Sociology," *Cultural Sociology*, 1 (1): 97–114.

Herrero, C. (2001), *Pablo Díez Fernández: Empresario Modelo*, Mexico City: Universidad Autónoma Metropolitana-Iztapalapa.

Hibino, H. (1992), "Cervecería Cuauhtémoc: A Case Study of Technological and Industrial Development in Mexico," *Mexican Studies/Estudios Mexicanos*, 8 (1): 23–43.

Howard, P. H. (2014), "Too Big to Ale: Globalization and Consolidation in the Beer Industry," in M. Patterson and N. H. Pullen, eds., *The Geography of Beer: Regions, Environment, and Societies*, 155–165, Dordrecht: Springer.

Jiménez, A. (2000), *Lugares de gozo, retozo, ahogo y desahogo en la Ciudad de México: Cantinas, pulquerías, hoteles de rato, sitios de prostitución, carceles*, Mexico City: Oceano de México.

Johnston, J. and S. Baumann (2010), *Foodies: Democracy and Distinction in the Gourmet Foodscape*, New York: Routledge.

"La Cervecería Cuauhtemoc" (1908), *El Mundo Ilustrado*, December 27, 1908: 844–845.

Mitchell, T. (2004), *Intoxicated Identities: Alcohol's Power in Mexican History and Culture*, New York: Routledge.

Morales, C. R. (n.d.), "La cerveza artesanal en México," *Mexico Desconocido*, www.mexicodesconocido.com.mx/cerveza-artesanal-mexico.html (accessed September 27, 2013).

Orozco y Berra, M. (1855), *Apendice al diccionario universal de historia y geografía*, Mexico City: J. M. Andrade y F. Escalante.

Paredes, A. (1976), *A Texas-Mexican Cancionero: Folk Songs of the Lower Border*, Austin: University of Texas Press.

Partlow, J. (2015), "Mexican Microbrewers Step Out of the Shadow of the Country's Beer Giants," *Washington Post*, July 13, 2015.

Payno, M. (1864), "Memoria sobre el maguey mexicano," *Boletin de la Sociedad Mexicana de Geografía y Estadística*, 10: 383–451.

Pilcher, J. M. (2012), *Planet Taco: A Global History of Mexican Food*, New York: Oxford University Press.

Recio, G. (2007), "El nacimiento de la industria cervecera en México, 1880–1910," in E. Sánchez Santiró, ed., *Cruda realidad: Producción, consumo y fiscalidad de las bebidas alcohólicas en México y América Latina, siglos xvii–xx*, 155–185, Mexico City: Instituto Mora.

Rojas Rivas, E., F. C. Viesca González, E. Espeitx Bernat and B. Quintero Salazar (2016), "El maguey, el pulque y las pulquerías de Toluca, Estado de México, ¿patrimonio gastronómico turístico?," *PASOS: Revista de Turismo y Patrimonio Cultural*, 14 (5): 1199–1215.

Ruiz, J. R. (2011), "Mexico," in G. Oliver, ed., *The Oxford Companion to Beer*, 583–584, New York: Oxford University Press.

Schultz, E. J. (2012), "How this Man Made Dos Equis a Most Interesting Marketing Story," *Advertising Age*, 83 (10), http://adage.com/article/behind-the-work/story-dos-equis-interesting-man-world/233112/.

Shepherd, G. M. (2011), *Neurogastronomy: How the Brain Creates Flavor and Why It Matters*, New York: Columbia University Press.

Taylor, W. B. (1979), *Drinking, Homicide, and Rebellion in Colonial Mexican Villages*, Stanford, CA: Stanford University Press.

Toxqui, Á. (2011), "Taverns and Their Influence on the Suburban Culture of Late-Nineteenth-Century Mexico City," in K. R. Hall, ed., *The Growth of Non-Western Cities: Primary and Secondary Urban Networking, c. 900–1900*, 241–269, Lanham, MD: Lexington Books.

Toxqui, Á. (2014), "Breadwinners or Entrepreneurs? Women's Involvement in the *Pulquería* World of Mexico City, 1850–1910," in G. Pierce and Á. Toxqui, eds., *Alcohol in Latin America: A Social and Cultural History*, 104–130, Tucson: University of Arizona Press.

Von Humboldt, A. (1836), *Examen político sobre la Isla de Cuba*, trans. Vicente González Arnao, vol. 2, Gerona: Imprenta de A. Oliva.

Wilson, T. M., ed. (2005), *Drinking Cultures: Alcohol and Identity*, Oxford: Berg.

Womack, J., Jr. (2012), *El trabajo en la Cervecería Moctezuma, 1908*, Mexico City: El Colegio de México.

11

Diffused Palates: The Evolution of Culinary Tastes of Jewish Mexicans Living in Israel

Paulette Kershenovich Schuster

AMILAT, Open University, The Hebrew University of Jerusalem

Introduction

A well-trained and refined palate can taste and differentiate ingredients and impeccably discern varying spectrums of flavors in dishes. There are many elements that are not taught or determined by physiology that can condition and shape our sense of taste and influence our palates. How we taste and perceive food is influenced by a host of factors such as the cultural relationship of individuals to food and eating, religious prescriptions, societal attitudes toward food within a specific context or setting, marketing trends, personal food habits, to name but a few. Food and foodways have been the subject of study by scholars in a variety of fields and provide an effective lens through which to broaden our own understanding of the complexities and nuances of our own food choices (among pioneering anthropologists see: Douglas 1966; Goody 1982; Harris 1986; Lévi-Strauss 1969).

Based on recent research work centered on the translational lives of Jewish Mexicans living in Israel, this chapter describes the evolution of taste among this migratory group. I suggest that food constitutes not only a link with their birth-nation but that it becomes a core element of their identity and being. In this chapter, I focus on how their culinary tastes have changed from their home country to their adopted homeland. I also examine the attitudes toward food, consumption, and everyday commensality. Particularly, I am concerned with understanding the role of commensality as a social device that is integral to the formation of personal and collective identities (Kerner, Chou and Warmind 2015).

This chapter is divided into four sections: (1) a brief introduction to the Jewish community in Mexico; (2) a description of their consumption of Mexican food in Mexico as part of their national diet and the ways in which it is mixed with Jewish cuisine; (3) an examination of their assimilation to the Israeli diet, their strategies of adaptation of Mexican food to Israeli dishes, and the creation of their own version of Mexican food in fusion with the Jewish Mexican food; and lastly (4) how Mexican food is seen in Israel by the local population, and the ways in which it has been appropriated by tourists as part of their own culinary repertoire. As I demonstrate, within this evolution of culinary tastes, certain aspects of belonging, solidarity, and acculturation become prominent.

Background to the Jewish community in Mexico

The first Jewish immigrants arrived in Mexico during the colonial period as *conversos* (converts or crypto-Jews) fleeing from the Spanish Inquisition. Later, during the late nineteenth century and the first half of the twentieth century, a wave of Sephardic Jews entered the country (Gojmande Backal 1987; Krause 1970; Sourasky 1965). The community was formally organized in 1912, just in time to welcome waves of *Ashkenazim* fleeing the Holocaust. Most Jews entered Mexico between the 1920s and 1950s. Three migratory periods with two major waves can be identified: 1876–1911, 1911–1934, and 1934–1950 (Kershenovich Schuster 2012). Today, the Jewish community in Mexico is relatively small in comparison to other communities in the diaspora. It numbers between 35,000–40,000, and it is an insular community with its own religious, social, and cultural institutions, located mostly in Mexico City, Tijuana, Cancún, Monterrey, and Guadalajara. Stavans (2011) wrote that a neutral local individual would probably say that the Jewish community lives *aislada* (isolated). From my experience this is not accurate, but Jews tend to keep to themselves out of fear of anti-Semitism and fear of assimilation.

The Mexican Jewish community is close-knit, highly functioning, organized, and centralized. It is divided into four main ethno-religious and cultural subgroups defined by their different places of national origin: Ashkenazim (Eastern European Jews), *Sephardim* (from the Iberian Peninsula, Turkey, and the Balkans), and two distinct Syrian Jewish subgroups (one Damascene, *Shamis*, and the other Aleppan, or *Halebis*). Within Jewish circles, each of these subgroups is known as a *comunidad* (community). Each Jewish community is distinct. These subgroups are divided even further into ten community sectors to which the majority of the Judeo-Mexican population is affiliated (Selingson

Berenfeld 1973: 203; Tribuna Israelita n.d.). Each community sector provides its members with a whole array of services, ranging from health and religious to recreational activities.[1] Each sector also has its own schools and synagogues, but members are free to attend any of their choosing. The Jewish community shares similar affinities and cultural commonalities, and encompasses the full spectrum of Jewish religious observance. The Ashkenazim and Sephardim tend to be traditional in their religious outlook (Roitman 1996), while the Shamis and Halebis tend to be more religious (Hamui 2005).[2] Most Jews in Mexico City live in suburban areas. Support of Israel has been a constant since prior to its establishment. Tourism and donations are high, and there are several Zionist youth groups that encourage *Aliyah* (immigration to Israel). Despite such support, the data from the Israel Central Bureau of Statistics reveals that immigration to Israel from Mexico is relatively low, especially when compared to other Latin American countries.[3]

Mexican vs. Israeli diet

To understand the culinary habits, food choices, eating and dietary patterns of the participants, I briefly describe the diets of each respective country. There are many similarities between the Mexican diet and the Israeli one, including the use of fresh seasonal vegetables and fruits, and the ways that food is prepared. Both countries emphasize sharing and enjoying meals as a collective. Eating patterns observed indicate that the food structure is the same in both countries. For example, consuming a light breakfast, a heavier afternoon meal, and a light dinner. However, the types of food consumed, combinations of ingredients, and times to eat are different.

Mexican diet

There is no typical or unified Mexican diet. Each state has its own regional dishes that are colorful and varied, much like the people of Mexico. There are many common elements that overlap, such as the consumption of *tortillas* (a flat type of round bread made out of maize *masa* or dough)[4] and other corn-derived foodstuffs, rice, beans, *chiles*, and salsas. The modern diet is derived from old ritual practices, farming methods, cooking techniques, community customs, and harvesting of ingredients.[5] The Mexican diet is mainly characterized by the consumption of corn, vegetables, fruits, and cereals.

Over time dietary habits, the food consumed, the way they are prepared, and commensality have changed. According to Arvizú Martínez, Polo Oteya and Shamah Levy (2013), in previous generations family members ate together, but today 46 percent of children eat while watching television. Additionally, in the last decades, the consumption of meat, dairy, fast food, and sweets has also increased. According to Garcia Uriguen (2012), the foodstuffs consumed in Mexico are mainly: tortillas, citrus (lemon, lime, oranges), vegetables, sugar, fruit, chicken, eggs, bread (including refined and sweet breads), beef, pork, green *chilis*, and beans. Other native staples include varieties of tomato (green and red), squashes, avocados, cacao, and vanilla. *Chiles* are an integral part of everyday food consumption. They are featured prominently in every meal and in snacks in the form of salsas, raw or pickled as accompaniments, and as a powder sprinkled over fruit, vegetables, corn kernels, ice cream, popcorn, potato chips, and popsicles.

Israeli diet

Israel is a country that comprises a myriad of cultures, nationalities, and flavors. This is reflected in a cuisine that encompasses dishes from a variety of Jewish ethnic traditional cuisines, influenced by their places of origin in Northern Africa, Europe, and the Middle East. In Israel, these cultures blend to form a melting pot. Their cuisines, which are influenced by kosher dietary laws, are divided into Mizrahi, Sephardi, and Ashkenazi. The Mizrahi cuisine reflects the converging preferences of Jews from Iraq, Syria, Lebanon, Yemen, Iran, Algeria, Morocco, Afghanistan, Bukhara (Uzbekistan), Egypt, the Berber communities, Kurdistan, Eastern Caucasus, and Georgia (Bahloul 1989: 85–95; Leichtman 2001: 247–272). It also shares a number of culinary practices with Palestinian Arabs (Dabbdoub Nasser 2001; Ranta and Mendel 2014: 1–24; Shihab 1993). The Sephardic cuisine includes dishes from Jews who arrived mainly from Spain, Greece, Bulgaria, and Portugal (Kershenovich Schuster 2015b).

Mizrahi and Sephardic dishes in Israel are often grouped together and not differentiated. This type of cuisine is light and fresh, concentrating on salads, stuffed vegetables and vine leaves, olive oil, lentils, fresh and dried fruits, herbs, nuts, and chickpeas. Lamb and beef are eaten more sparingly. Rice is a staple of the Mizrahi diet, as are many varieties of flatbread such as *lafah* (taboon flatbread or Iraqi pita), *pita*, and *malawach* (a pan-fried flat puff pastry). There are intragroup variations and modifications. The Ashkenazi cuisine incorporates dishes eaten by Jews from Eastern and central Europe such as Poland, Lithuania,

Russia, Hungary, and Germany. The Ashkenazi Jewish diet is largely based on potatoes, beets, brisket, chicken, and cabbage in various preparations (Bernstein and Carmeli 2004).

For a long time, food preferences among Israelis followed ethnic separations and political divides (Dahan Kalev 2001; Khazzoom 2008; Smooha 2008). However, today, there is a growing acceptance of culinary diversity that suggests that Israelis are more willing to sample newer culinary traditions and even adopt certain new elements. When eaten, these ancestral ethnic cuisines tend to be consumed side by side in what I call "parallel cuisines" and "parallel eating," rather than food fusion.[6] Food fusion occurs in Israeli cuisine but mainly with foreign cuisines. Some fusions become successful, for example, sushi with hot local peppers and Israeli spices; and hamburgers complemented with local cheeses and sauces. However, some experiments fail, for example, an Israeli salad wrapped in a tortilla. These dishes are usually served in chef-owned or chef-run restaurants that are not kosher (on account of the combination of meat and dairy, and because they open throughout the weekend, including Saturdays). Jewish dishes are often served together with other Middle Eastern foods such as red or green Yemenite *szug* (a hot chili pepper sauce), which is often added to falafel, as a dip for vegetables, or doused on top of hummus and spread over eggs or *chraime* (a spicy Moroccan fish with tomatoes). This type of "parallel eating" and "parallel cuisine" is quite common.

Israeli cuisine, like other Middle Eastern and Mediterranean cuisines, is based on a few key ingredients such as lamb, chicken, eggplants, rice, *burghul* (cracked wheat), olives, yogurt, whole grains, vegetables, legumes, pulses, olive oil, and salty cheeses (Dweck and Cohen 2007; South and Jermyn 2006: 122).[7] Traditional Jewish dishes largely follow the cuisines of the countries of origin, except for the separation between dishes containing meat and those containing dairy ingredients (according to kosher dietary laws). Generally speaking, the modern Israeli weekday diet is based on fresh vegetables, fruits, a wide variety of olives and pickled vegetables (eggplant, carrot, cabbage, cauliflower, squash, and lemon), and dairy products (Golden 2005: 181–199; Gur 2008; Ziv 2013). Eggs, bread, chicken or chicken *schnitzel* (thinned chicken steak, usually by pounding with a meat tenderizer, that is breaded and fried in oil) are also consumed on a regular basis. Beef, lamb, and fish are not central to everyday consumption, but this seems to be changing. These dishes are usually reserved for the *Shabbath* and festival and celebratory meals. Weekday food in Israel also consists of a mix of styles: mostly Israeli and traditional Jewish dishes mixed with international dishes (for instance, sushi, pizza, hamburgers, or dim sum). Typical dishes consumed on a regular basis include *memula'im* (stuffed vegetables), *falafel* (fried balls or patties of spiced, mashed chickpeas or fava beans often served in a pita, along with pickles), *shwarma* (shredded lamb, turkey, or chicken dish), *hummus, tahini* (sesame spread), *shakshuka* (poached eggs cooked in a spicy tomato and hot pepper sauce), and *couscous* (cooked semolina in an aromatic broth).

Jewish consumption of Mexican food in Mexico

In general, when Jews first arrived in Mexico they ate the food that they were familiar with back home. For the Ashkenazi Jews that meant, for example, *Borscht* soup (sour soup made from beetroot and dolloped with sour cream), for the Sephardim it was *Sofrito* (aromatic ingredients cut into small pieces and sautéed), and for the Halebis and Shamis various types of *kipe* (dumplings made with burghul, stuffed with minced meat and spices). As time wore on, Mexican dishes gradually entered their cuisines. Second and third generations have adopted modern Mexican patterns of eating as well as Mexican foods. The tortilla has become an integral component, present almost at every meal, along with refried beans and Mexican red rice. Today, eating patterns of fourth and fifth generation Mexican Jews vary according to each Jewish community. The community of Jews from Syria tends to consume its own ethnic cuisine, whereas Ashkenazim or Sephardim often eat the food of the Syrian Jews as well as their own. This divide is indicative not of the different palates but of cultural distinctions and levels of religiosity. In Mexico, traditional family meals are eaten on *Shabbat* and are often served in a special dining area used only for Shabbat, holidays, and other festive occasions.

All communities eat Mexican food, in one way or another. Depending on their religiosity, Mexican food plays a lesser or greater role in their diets. As mentioned above, the Jewish community in Mexico encompasses many forms of religious observance, and some Jews participate in kosher dining while others do not. This determines their consumption of certain local flavors that are not kosher; for example, *cochinita pibil* (slow-baked pork) or *chicharron* (fried pork skin), or the extent to which they adapt these dishes to meet their dietary observances. Jews eat Mexican food in different ways. They may eat Mexican dishes, alongside Jewish dishes. For example, they start a Shabbath meal with matzah ball soup, followed by Mexican red rice and beans as side dishes to *carne a la tampiqueña* (Tampico-style grilled beef tenderloins). In some cases, but not necessarily, the cut of meat will be kosher. It may be topped with shredded *queso fresco* (hard white cheese) and/or roasted onions and poblano/serrano peppers. Some may leave the cheese out and replace it with salsa to avoid mixing dairy and meat.

In other cases, Jews mix Mexican ingredients with Jewish cuisine, in accordance with their place of origin. For example, Syrian Jews eat okra with tomatoes (traditional Jewish dish) combining it with tamarind and chile ancho (Mexican elements), or they eat kipe (traditional Syrian/Jewish dish) doused with green salsa (Mexican element). In the Ashkenazi community, *gefilte fish* (traditional Jewish dish) *a la Veracruzana* (Veracruz-style) is quite common, and is routinely enjoyed on holidays and for Sabbath meals.[8] Jews may also add typical Mexican spices to chicken soup. Kosher consumers also enjoy Mexican staples, like *antojitos* (Mexican appetizers) such as *sopes* (small deep-fried disks of cornmeal dough), *chalupas* (cups of deep-fried cornmeal dough), tostadas—which may be topped or filled with Mexican ingredients such as *nopales* (cacti), refried beans, salsas, and guacamole—or heavier main dishes such as *quesadillas* (folded corn tortillas that are stuffed with cheese then pan- or deep-fried), *flautas* (rolled and deep-fried tortillas), and *enchiladas* (rolled corn tortillas covered and baked with salsa). The latter may be filled with a variety of ingredients, including different meats, cheese, beans, potatoes, vegetables, or their combinations. Many kosher restaurants prepare Mexican dishes observing dietary laws, and even nonkosher restaurants include kosher cuts of meat serrated with a special knife and cooked in a kosher designated area. However, more often than not, these kosher cuts are rendered not kosher because they are served in dishes or prepared in areas that are not kosher.

Mexican food in Israel

The attitudes toward Mexican food in Israel have changed over the years. Historically, Israelis have not always been keen on Mexican food, and have viewed it with some skepticism. For example, once I was told that corn tortillas smelled and tasted like soap. Mexican food is often seen as just spicy street food. More often than not, when restaurants have sought to recreate Mexican dishes in Israel they have adapted popular dishes to comply with the Israeli palate, resulting in disastrous combinations such as *tacos* made out of cold whole wheat tortillas filled with chicken *schnitzel*, diced tomatoes, and doused with tahini, or their menus are filled with a mishmash of ingredients. Mexican restaurants cater mainly to foreigners living in Israel, American tourists, and to the small population of Mexicans (Jews and non-Jews alike) that call Israel home. These restaurants form part of gastronomic tourism. This type of tourism also incorporates social tourism (Toth Stub 2017). In both cases, women open up their homes to showcase their cooking practices. The former do it for commercial gain, and the latter through charities seeking to benefit the community where they live. Social tourism is further enhanced by participation in cooking courses and/or taking tours to companies that manufacture traditional products. Mexican women have chosen to start catering businesses to offer authentic Mexican food to Mexicans and Israelis, and often hold workshops on Mexican food and cooking. Due to the unavailability of prime Mexican ingredients, some of these women, and some men, have begun growing their own *tomatillos* (small green tomatoes used for green salsa), *jalapeños* and *poblano* peppers at home, and make tortillas for commercial profit.

Recently, a handful of Mexican dishes have reached the Israeli culinary culture/market: tacos, enchiladas, quesadillas, tostadas, and *churros* (deep-fried pastry dough fingers served with cinnamon, sugar, and caramel or chocolate). Tacos are often *al pastor* (chicken or beef marinated in *achiote* sauce, served with pineapple and other fixings), *tinga* (shredded chicken with sautéed onions in *chipotle* [smoked jalapeño] tomato sauce), *carne asada* (grilled strips of meat), fish (marinated in various sauces and then grilled), or *barbacoa* (slow-roasted beef); enchiladas (which are either green or red) are filled with chicken, beef, or mushrooms; and tostadas are either topped with beans, lettuce, tomatoes, and cheese or chicken. Other popular items are rather Tex-Mex than Mexican: *nachos* (tortilla chips topped with refried beans, guacamole, chopped tomato and onion salsa, jalapeños, melted cheese, and a slab of sour cream), *burritos* (a large wheat flour tortilla stuffed with different fillings), *chimichangas* (deep-fried burritos), and *chili con carne* (a stew of beans and ground beef meat). I found that, in the different restaurants I surveyed, these dishes were served as variations of the same meal. When I asked the owners why this was the case, they all basically responded that they were afraid to deviate from the recipe. As one restaurateur told me: "Why tamper with a given formula? This works, so let's stick with it." Experience has shown that when restaurants have deviated from this tried and true formula, they have quickly failed and closed. Ironically, but not surprising, none of the Mexican restaurants are owned or run by Mexicans but by Israelis who have returned to Israel after living in Mexico for various lengths of time. These Israelis fell in love with the cuisine of their partners—who they also imported. Some of these dishes are sold in Israel as authentic Mexican food, often along with Tex-Mex dishes that are better known by local Israelis and tourists. American tourists especially ache for memories of home and often eat Mexican food as a strategy to reconnect to their home country and to distance themselves from local Israeli cuisine, which they find too spicy. In this sense, they have appropriated these meals as symbols of home. When Mexicans in Israel serve other dishes at events, or in their homes, Israelis get confused. As one wrote in Hebrew in a Facebook group: "What are all these special things? Could you please explain what they are so we could understand?" The repetition of the same type of Mexican dishes throughout Israel is a strategy to ensure business success. Restaurateurs have adapted these dishes to fit mainly the Israeli and American palates. They have renewed Mexican food with Israeli flavors. For example, the hot sauces used for marinating and certain side dishes do not appeal to the Mexican palate. Rather, they contain ingredients such as coriander, garlic, green local peppers, and cumin, which are similar to the ingredients in szug.

These restaurants do not target the Mexican population, but Mexicans do not avoid them. They frequent these establishments as a way to relive memories of home, to gather socially, and to reconnect with other compatriots. As they dine, through commensality, they recreate a sense of belonging. Memory, as we know, is often strongly tied to foods. It is a shared emotion among many ethnic groups, and cuts across the generational divide. It is not about place but an internal yearning that seeks to affirm a connection that anchors unspoken feelings into something concrete and tangible. This is the power of food.

Commensality, memory, and identity

During my research I have found multiple meanings of food and the part it plays as a marker of personal and social identity. The cultural production of food took place in two settings: in the private domain, it was performed within the family at home; in the public sphere, it took the form of consumption in restaurants, outdoor picnics, barbecues, and different social events. Food, for some of the participants in this study, was seen as a vehicle and gesture of love, and cooking was seen as necessary to communicate those emotions. As one informant told me: "Food is about love and family. Coming together to enjoy a meal and talk about our day. To share what is important." For others, it was about connecting with people and affirming their identity not just as Mexican Jews in Israel but also as something that is more complex and multilayered.

Food constitutes not only a link with their birth-nation and with memories of home, it becomes a core element of their present identity and being. They use food simultaneously as an anchor to the past, a link to the present, and a tool to teach their children about their Jewish-Mexican heritage. Their attitude varies depending on the place of consumption and the company they enjoy. Within their families dishes are more elaborate, and family recipes are used to inscribe histories and stories. For them, it is important to sit around the table in everyday commensality sharing the food they love. In contrast, with friends, simple-known dishes that appeal to Mexicans and Israelis alike are consumed. In Mexican-only gatherings the dishes are spicier. In these cases, dishes are used to display Mexican culinary traditions and to highlight flavors that Israelis are not accustomed to.

Among many participants, when they recalled memories about Mexico or their families, their narratives shifted to what I call "food tales": vignettes about the time surrounding events or gatherings where food or cooking was involved. In these stories, the centrality of food preparation and consumption, in both a religious and secular context, was highlighted.[9] For example, the memory of which food was served at the circumcision ceremonies of sons, wedding celebrations, and Sabbath dinners with grandparents.

> When I cook for my children, I think of my *Bobe* [grandmother in Yiddish] and what she cooked for me. The *gefilte* fish, the *kugel* [traditional Ashkenazi noodle savory or sweet dish], *kroitlaj* [stuffed cabbage, traditional Jewish Ashkenazi dish] filled with tamarind [Mexican sweet and sour fruit], meat and stuffed quail. It brings me great joy and sadness at the same time. She may no longer be with us but her recipes will live in my heart forever and hopefully my children will continue cooking these family recipes as well. Of course, in a different way. Here in Israel, I don't make quail, I use chicken, which my kids prefer and that's ok as long as the core stays the same.

This is a revealing expression of longing. Nostalgia and longing are powerful sentiments that evoke feelings of belonging, solidarity, and ambivalence. The culinary chats I was involved in revealed that the most important elements in cooking is the special *toque* (touch) of love and *gusto* (taste, *tam*) that one brings to a dish (Abarca 2001: 133; Schreibman Walter 2018). The language of cooking remains the same while creating new fusion dishes and recreating old ones seeking to appeal to nostalgic feelings and to recapture memory.

I found in conversations with men and women that there were many instances in which men did the cooking on Friday afternoons for the *Kabbalat Sabbath* (receiving of the Sabbath) dinner. This was striking since these men were secular rather than religious. In these cases, the food served was purely Mexican, excluding other traditional Jewish or Israeli dishes. During the week, Mexican Jews ate Israeli food liberally outside the home. They ate shwarma, falafel, hummus, and other street food, but did not cook at home unless it was mixed with Mexican ingredients. For example, combining shwarma with different salsas, adding jalapeños to falafel, or adding refried beans to hummus. These culinary concoctions and combinations are just some examples of the food fusion between Israeli local cuisine and Mexican fare that takes place within Jewish religious traditions, or what I call the "Mexicanization of Israeli foodways" (Kershenovich Schuster 2015b). At other times, these men and women tried to recreate Mexican signature restaurant dishes like baguette sandwiches with shredded meat, parmesan cheese, and *chiles toreados* (roasted jalapeños and onions), or chipotle brownies.

At home, food is served as a way for women and men to teach their children about their Mexican and Jewish heritage while transmitting unspoken cultural codes. Commensality functions as an intersecting bond that unites their own cultural heritage and religious traditions, an expansive conceptualization of identity, and reveals the ambivalences located in everyday moments centered on food. Home-cooked meals are prepared to keep nostalgia at bay, to feel connected to homes and families. The food is as unadulterated and authentic as it can be. Perhaps some cooks adapt local ingredients when it is necessary to do so. For Jews, cooking their native cuisines is a vehicle for remembrance and memories that have been coded and laced with spices and flavors from a home they have for the most part willingly abandoned. They often combine both cuisines into a single dish, eating shakshuka with jalapeños, or falafel with salsa. As one woman in her thirties commented, "I add my *salsita* to everything; if I don't, it doesn't taste good."

At public spaces, it is commensality that takes on a central role over the food. Women gather with friends to perform social acts and to feel connected to their homeland, language, and culture. Food in this case is important but not central. Food is the reason to gather, but what is served is not crucial. Meetings in restaurants, potlucks or "bring-a-dish" picnics and barbecues are popular. During these reunions, commensality constitutes a central practice in social mapping. The participants form a social group that creates and strengthens relationships and bonds. At the same time, it also sets boundaries, demarcating in-group and out-group membership according to religious beliefs or societal norms. As they share meals, food becomes a tool for communication and of bonding over shared experiences.

For immigrants, food is an important element in the maintenance of personal ties with their home country and a cohesive factor in the construction of a new identity in Israel, their adopted homeland. Food practices encode tacit information and nonverbal cues that are integral parts of an individual's relationship with different social groups. Their lives revolve around food as a reminder of a lost home: "I miss Mexico, I miss my family and I miss the food … oh the food. I can't get decent *chiles* here. I bring them over as much as I can." This duality of self-identification and cultural pride is linked to food. Food becomes a potent marker of identity. Nostalgic memories of a home that is no longer there, but yet remains, become reminders of another place and time, which continues to shape present tastes, flavors, and modes of cooking. In essence, blending their Mexican past with their Israeli future.

Conclusion: Tasting home in Israel

Mexican food has become a global fashion but is also locally shared and shaped. In the case of Jewish Mexicans living in Israel, Mexican food is more than a social marker; it is a source of identity anchored in the replication of loved dishes, in gathering for special events, and in its weekly consumption. I suggest that food constitutes not only a link to their birth-nation but that it becomes a central locus of their identity and being. Despite the adaptation of Mexican food to local flavors, its culinary taste has changed little. Mexicans still long for hot sauces, the smell of fresh-baked *bolillos* (Mexican French-style rolls), and the tastes of home. To quench their desire, they cook Mexican dishes at home on a regular basis, adapting local ingredients to fit and recreate their recipes. Food for them is about the sense of family, the imagination of continuity in food culture, of nostalgia, love, and everyday commensality, and is also a fundamental part of their collective identity as Mexican-Israelis or as Jewish Mexicans in Israel.

Notes

1 Jewish organizations provide legal advice, financial and welfare assistance, educational funds, independent ambulance services, social groups and activities, anti-kidnapping and extortion response groups, and burial and cemetery services.

2 For more on the religious spectrum and how Jews define and classify themselves, see DellaPergola and Lerner (1995); and Alduncin and Asociados (2000).

3 Since 1948, fewer than 6,000 Jews have migrated from Mexico to Israel. See the Israel Central Bureau of Statistics (n.d.).

4 Tortillas vary in size and color depending on the type of corn used (white, yellow, blue or violet, and/or red corn). They may be small (2 inches in diameter) or large (as wide as 10 inches).

5 Such as *milpas* (rotating fields of corn and other crops) and *chinampas* (man-made farming islets in lake areas); and cooking techniques like *nixtamalización* (lime-hulling maize) (UNESCO n.d.).

6 I call "parallel eating" the eating of different types of foodstuffs in parallel form—such as meat, tortillas, and chicken—but not together on the same plate. In addition, I call "parallel cuisine" the serving of various types of national cuisines side by side but not on the same plate. For example, serving sushi as an appetizer, Indian chicken curry and white rice as the main and side dishes, and Israeli *halva* (sesame seed candy) as dessert. Food fusion, a process characteristic of the immigrant experience, refers to the trend of combining food elements from more than one culture in the same dish. For example, making Swedish meatballs with Mexican hot peppers or local ground meat.

7 On food in the Middle East, see Heine 2004; Zubaida and Rapper 1994; and Roden 1994. For a general background on the Middle East and its Jewish populations, see Cleveland 2004; and Simon, Laskier and Reguer 2003.

8 The origin of this popular dish is unknown. It is based on *Huachinango a la veracruzana* (Veracruz-style red snapper). Virtually every Ashkenazi household eats a variation of this dish either at Rosh Hashanah, Passover, or Shabbat meals (Lieberman 2017). As an anecdote, when I was growing-up my grandmother prepared this dish and I naïvely thought it was a passed down recipe that only my family enjoyed. When I grew older and wiser, I discovered that this dish is quite common among Mexican Ashkenazi households.

9 I have addressed the sacredness of food and how religious observance influences even the mundane meals and imbues them with symbolic meanings (Schuster 2002). For Jewish cooking and eating patterns, see Kershenovich Schuster 2015a, 2015c.

References

Abarca, M. E. (2001), "Los chilaquiles de mi 'ama: The Language of Everyday Cooking," in S. A. Inness, ed., *Pilaf, Pozole and Pad Thai: American Women and Ethnic Food*, 119–144, Cambridge, MA: University of Massachusetts Press.

Alduncin and Asociados (2000), "Estudio sociodemográfico de la comunidad judía de México: a solicitud del Comité Central de la comunidad judía en México," Unpublished, México City.

Arvizú Martínez, O., E. Polo Oteya and T. Shamah Levy, eds. (2013), *Qué y cómo comemos los mexicanos: Consumo de alimentos en la población urbana*, Mexico City: FUNSALUD and INSP.

Bahloul, J. (1989), "From a Muslim Banquet to a Jewish Seder: Foodways and Ethnicity among North African Jews," in M. Cohen and A. Udovitch, eds., *Jews among Arans: Contacts and Boundaries*, 85–95, Princeton, NJ: The Darwin Press.

Bernstein, J. and Y. S. Carmeli (2004), "Food for Thought: The Dining Table and Identity Construction among Jewish Immigrants from the Former Soviet Union in Israel," in Y. S. Carmeli and K. Applbaum, eds., *Consumption and Market Society in Israel*, 71–95, New York: Bloomsbury Academic.

Cleveland, W. (2004), *A History of the Modern Middle East*, Boulder, CO: Westview Press.

Dabbdoub Nasser, C. (2001), *Classic Palestinian Cookery*, London: Saqi Books.

Dahan Kalev, H. (2001), "Tensions in Israeli Feminism: The Muzrahi-Ashkenazi Rift," *Women's Studies International Forum*, 24 (6): 1–16.

DellaPergola, S. and S. Lerner (1995), *La población judía de México. Perfil demográfico, social y cultural*, Jerusalem: Hebrew University Press, Avraham Harman Institute of Contemporary Judaism and COLMEX.

Douglas, M. (1966), *Purity and Danger: An Analysis of Concepts of Pollution and Taboo*, London: Routledge and Kegan Paul.

Dweck, P. and M. Cohen (2007), *Aromas of Aleppo: The Legendary Cuisine of Syrian Jews*, New York: Ecco.

Garcia Uriguen, P. (2012), *La alimentación de los Mexicanos: Cambios sociales y económicos, y su impacto en los hábitos alimenticios*, Mexico City: Canacintra.

Gojmande Backal, A. (1987), "Los conversos en el México Colonial," in *Jornadas Culturales—La presencia judía en México: Memorias*, Mexico City: UNAM, Tribuna Israelita and Multibanco Mercantil de México, SNC, México.

Golden, D. (2005), "Nourishing the Nation: The Uses of Food in an Israeli Kindergarten," *Food and Foodways*, 13 (3): 181–199.

Goody, J. (1982), *Cooking, Cuisine and Class: A Study in Comparative Sociology*, New York: Cambridge University Press.

Gur, J. (2008), *The Book of New Israeli Food*, Tel Aviv: Al Hashulchan.

Hamui, H. L. (2005), *Transformaciones en la religiosidad de los judíos en México: Tradición, ortodoxia y fundamentalismo en la modernidad tardía*, Mexico City: Grupo Noriega Editores.

Harris, M. (1986), *Good to Eat: Riddles of Food and Culture*, London: Allen and Unwin.

Heine, P. (2004), *Food Culture in the Near East, Middle East, and North Africa*, Westport, CT: Greenwood Press.

Israel Central Bureau of Statistics (n.d.), www.cbs.gov.il (accessed November 28, 2017).

Kerner, S., C. Chou and M. Warmind, eds. (2015), *From Everyday Food to Feast*, London: Bloomsbury.

Kershenovich Schuster, P. (2012), *The Syrian Jewish Community in Mexico City in a Comparative Context: Between a Rock and a Hard Place*, Saarbrücken: LAP Lambert Academic Publishing.

Kershenovich Schuster, P. (2015a), "A Tapestry of Tastes: Jewish Women of Syrian Descent and Their Cooking in Mexico and Israel," in A. Helman, ed., *Jews and their Foodways: Studies in Contemporary Jewry*, 160–176, Oxford: Oxford University Press.

Kershenovich Schuster, P. (2015b), "Habaneros and Shwarma: Jewish Mexicans in Israel as a Transnational Community," *Scripta Instituti Donneriani Aboensis*, 26: 281–302.

Kershenovich Schuster, P. (2015c), "Jewish Food in Mexico: Reflection of a Community's History, Culture, and Values," *Religion and Food*, 15 (1): 58–79.

Khazzoom, A. (2008), *Shifting Ethnic Boundaries and Inequality in Israel*, Stanford, CA: Stanford University Press.

Krause Azen, C. (1970), "The Jews in Mexico: A History with Special Emphasis on the Period from 1857 to 1930," PhD dissertation, University of Pittsburgh, Pittsburgh.

Leichtman, M. (2001), "The Differential Construction of Ethnicity: The Case of Egyptian and Moroccan Immigrants in Israel," *Identity*, 1 (3): 247–272.

Lévi-Strauss, C. (1969), *The Raw and the Cooked*, New York: Harper & Row.

Lieberman, H. (2017), "Mexican-Jewish Food Is the Obsession You Need Right Now," *Tablet*, www.tabletmag.com/scroll/252114/mexican-jewish-food-pati-jinich-gefilte-fish-veracruzana (accessed December 28, 2017).

Ranta, R. and Y. Mendel (2014), "Consuming Palestine: Palestine and Palestinians in Israeli Food Culture," *Ethnicities*, 14 (3): 412–435.

Roden, C. (1994), "Jewish Foods in the Middle East," in S. Zubaida and R. Rapper, eds., *Culinary Cultures of the Middle East*, 63–73, London: I.B. Tauris.

Roitman, D. (1996), "Jewish Identification among Young Mexican Jews," MA Thesis, Bar-Ilan University, Department of Sociology, Ramat Gan.

Schreibman Walter, A. (2018), "Getting Back to Basics in the Kitchen," *Tablet*, www.tabletmag.com/jewish-life-and-religion/253228/how-to-cook-like-a-jewish-mother (accessed February 5, 2018).

Schuster, P. (2002), "Evoking the Essence of the Divine: The Construction of Identity through Food in the Syrian Jewish Community in Mexico," *Nashim: A Journal of Jewish Women's Studies and Gender Issues*, 5 (5763): 105–128.

Seligson Berenfeld, S. (1973), "Los judíos en México: Un estudio preliminar," BA Thesis, INAH, Mexico City.

Shihab, A. (1993), *A Taste of Palestine: Menus and Memories*, Texas, CO: Corona Publishing.

Simon, R. S., M. M. Laskier and S. Raguer, eds. (2003), *The Jews of the Middle East and North Africa in Modern Times*, New York: Columbia University Press.

Smooha, S. (2008), "The Mass Immigrations to Israel: A Comparison of the Failure of Mizrahi Immigrants of the 1950s with the Success of the Russian Immigrants of the 1990s," *Journal of Israeli History*, 27 (1): 1–27.

Sourasky, L. (1965), *Historia de la comunidad israelita de México, 1917–1942*, Mexico City: Moderna Pintel.

South, C. and L. Jermyn (2006), *Cultures of the World*, Syria: Bedrock of Civilizations, New York: Times Editions.

Stavans, I. (2011), *Return to Centro Histórico: A Mexican Jew Looks for His Roots*, New Brunswick, NJ: Rutgers University Press.

Toth Stub, S. (2017), "Next Stop on the Tour: The Living Room," www.tabletmag.com/jewish-life-and-religion/227814/tours-by-women-of-israeli-homes (accessed March 27, 2017).

Tribuna Israelita (The Jewish public relations agency in Mexico City) (n.d.), www.tribuna.org.mx/presencia-judia-en-mexico.html (accessed November 20, 2017).

UNESCO (United Nations Educational, Scientific and Cultural Organization) (n.d.), "Traditional Mexican cuisine—ancestral, ongoing community culture, the Michoacán paradigm," https://ich.unesco.org/en/RL/tradition-mexican-cuisine-ancestral-ongoing-community-culture-the-michoacan-paradigm-00400 (accessed December 22, 2017).

Ziv, O. (2013), *Cook in Israel: Home Cooking Inspiration*. Self Published.

Zubaida, S. and R. Rapper, eds. (1994), *Culinary Cultures of the Middle East*, London: I.B. Tauris.

12

Defining Sanitized Taste and Culinary Tourism in Cozumel, Mexico

Christine Vassallo-Oby
State University of New York–Albany

The island is forced to comply with the cruise lines; without it, we'd go hungry.

Cozumel store owner, field interview (January 2013)

Sanitized taste in the context of tourism studies

Sanitized taste is synonymous with North American palates, especially when describing the country's relationship with Mexican cuisine. For example, the advent of the *chimichanga* (a deep-fried *burrito*) on Texas soil embodies a mimicry of traditional Mexican cooking. The burrito is transformed into a more craveable, high-fat version of itself that caters to a primarily North American penchant for food that is salty, fatty, and sweet—a process similar to the transformation from steamed or baked potato to French fries. People from the United States crave the sanitization of Mexican food so much that Tex-Mex (an entire food genre), Taco Bell (a popular fast-food restaurant with 7,000 locations worldwide), and even "taco seasoning" were all created to support this uniquely American habit of enshrining ground beef, shredded cheddar cheese, iceberg lettuce, and sour cream into *tacos*, burritos, and *quesadillas*. Sanitization of traditional or indigenous foodstuffs from Mexico (e.g., corn *tortillas*) represents the transformation of particular foods or ingredients into Americanized versions.

In the context of tourism, sanitized taste describes how corporate factors influence tourists' sensory perceptions and toured environments. This corporate influence is done mindfully, to meet economic ends—laying aside personal subjectivity, or one's literal taste—as food and drink, as well as environments and spaces, are manipulated to fit the tourist consumption of the toured environment. International and national governments and tourism developers, multinational and national corporations, and the global tourism industry all play a role in creating and managing tourist taste. Corporate influence on tourist taste is exemplified in the context of cruise ship mass tourism both on board and on land. On board, there is an overabundance of all-inclusive food offered to tourists that is displayed in the form of buffets and, more recently, satellite locations of land-based restaurants including Johnny Rockets and Starbucks. On land, Cozumel is the most important cruise ship destination in Mexico and one of the most economically vital port-of-call communities in the world (Preble 2014). Despite the presence of vibrant local culinary options in Cozumel, multinational restaurant chains—such as Hooters and Señor Frogs—reign in and around the three cruise ship piers where over 3.3 million tourists disembark annually. This appears to be by design, as the cruise ship industry contracts with local franchises on and around the pier, effectively setting the limits of any visitor's culinary experience and therefore taste. Theoretically, we can begin to contextualize how tourist taste is influenced by entities beyond the self with both the tourist gaze (Urry and Larsen 2011) and purposefully built environments intended for cruise ship tourist consumption (Preble 2014).

The tourist gaze provides a conceptual framework that describes how print advertising attracts tourists to explore certain localities based on their sensory perception of written and visual images. For example, an image of a deserted beach with clean, white sand and a hammock gently swinging in the breeze beckons the viewer to picture themselves laying in the hammock, feeling the warm breeze on their skin. In viewing this image and simulating themselves in this particular environment, (potential) tourists escape their hectic lives in favor of solitude and a spiritual connection with the surrounding, pristine setting. This visceral reaction that tourism marketers generate through advertising is only one example of how the industry affects tourist taste.

The existence of physical spaces in mass tourism zones that are built exclusively for tourists to enjoy further captures how tourism affects tourist taste. This is especially evident on the cruise ship piers that dot each port-of-call community developed by cruise ship tourism, including in the Mexican Caribbean, where Royal Caribbean Ltd. and Carnival Corporation & PLC[1] have built three piers in the port-of-call community of Cozumel.[2] Each pier runs from the gangway where cruise ship tourists disembark to the point of entry on the island's soil. These piers are lined with duty-free shopping containing many of the same objects also found not only inland but at each port of call in the Caribbean (e.g., Diamonds International jewelry, Del Sol apparel, and Los Cinco Soles, which sells mass-acquired Mexican folk art and handicrafts). Consumer taste is bounded by the lack of choice offered to cruise ship tourists on each pier. The US-based cruise ship industry, which includes the ships owned and operated by Royal Caribbean Ltd. and Carnival Corporation & PLC, manipulates the physical space where each cruise ship docks and cruise ship tourists disembark. Maps of the port of call are delivered to each stateroom before disembarkation, each illustrating an industry-curated list of stores and restaurants in the area. Royal Caribbean's *Cozumel Port and Shopping Map* (2008) states in bold text, "Only the stores listed on this map are guaranteed by the cruise line." Royal Caribbean's *Port Shopping Map* (2016) boasts, "Shop with confidence. All merchants on this map have been carefully selected; each offers a 30-day guarantee to cruise line guests."[3] This map identifies which spaces are acceptable, recommended, and safe for cruise ship tourists while ashore. Tourists' experiences on shore are therefore structured by this map, as its language signifies every other space beyond the map, or a retailer or restaurant that is *not* listed, as potentially untrustworthy and unsafe for cruise ship tourists to patronize. Spaces outside or beyond this map are not for tourists to consume.

Additionally, only cruise ship tourists can access many of these spaces; the local community is excluded entry from piers and cruise ship shopping malls (e.g., International Pier, Punta Langosta pier, and Puerta Maya shopping destination and pier). This lack of local presence further complicates tourist taste as the distance between host and guest is purposefully kept in sharp contrast to benefit businesses that are not locally owned or operated. We can categorize tourist taste in the context of cruise ship mass tourism as purposefully sanitized. Sanitized taste is defined as both the tourist imaginary and actual sensory perception of the space they temporarily inhabit as guests. The tourist gaze influences their desires, expectations, and ideas of where they are traveling before and during travel. This type of taste is illusory as it is based on a preconceived set of ideas about their destination. In Cozumel, cruise ship tourists communicated their preconceived perceptions of safety, conviviality of the local population, and excitement to "let loose" (i.e., indulging in alcohol or reveling in not being responsible for anything or anyone other than themselves while on vacation) before they disembarked. During their tour of Cozumel, many cruise ship tourists can be seen "tasting" from menus and relishing in built environments that are decidedly familiar to their lives at home. In the context of cruise ship mass tourism, sanitized taste becomes part of the tourist bubble (Judd and Fainstein 1999) and integral to its functioning. The toured space simply becomes part of the background landscape, as the storefronts and restaurants, and therefore tastes, are replicated in each port of call. This tourist bubble discourages local presence, metaphorically and physically, as taste is centered on the familiar to ultimately comfort cruise ship tourists and, more cynically, to keep tourism revenues in the hands of foreign entities.

Entertainment restaurants: Consciously sanitized culinary options

Royal Caribbean's *Port and Shopping Guide* (2008) advertises Jimmy Buffet's Margaritaville on its front page. On the map illustrated on the next page, Margaritaville is described as, "Where the Wet and Wild Fun Never Stops: Live music, tropical drinks, great food; Try their water trampoline! Not to be Missed, New in Town: Stop at the Trading Post for exclusive Jimmy Buffett merchandise." A few years later in 2016, Royal Caribbean has updated the description; below the heading "Local Flavors," it reads, "It's always 5:00 here. Located right outside the cruise terminal at the International Pier. Unlimited fun and a lifetime of memories. Lunch and dine on delicious Caribbean and island cuisine while grooving." The restaurant chain operates "over thirty locations around the world" and is described on their website as "a state of mind." Founded on the singer/songwriter Jimmy Buffett's success with the 1970s song, "Margaritaville," the restaurant chain functions as a corporate empire, with its website advertising merchandise (clothing, shoes, hats), branded beach resorts and casinos, group tours and travel, and an online food retail site selling salsa, marinades, and frozen seafood. Each of these business ventures is characterized similarly with images of abandoned tropical beaches, empty hammocks swaying in the breeze, and icy, umbrella drinks. These images portray paradise, which is textually exemplified on the Margaritaville's website by the conglomerate's motto:

> There is a beach and a thatched roof bar perched on the edge of the turquoise sea where you can always find a bar stool. There are lots of lies and loads of stories. It's a comical concoction that blends together like tequila, salt, and limes. Where is Margaritaville? It's in your mind. (Margaritaville 2018)

Margaritaville builds a fantasy of paradise, physically in each of their storefronts and emotionally for their customers (tourists) by constructing an idealized version of taste as a "state of mind," a boozy paradise with all the comforts of home.

Margaritaville and other similarly advertised "entertainment restaurants" catering to international tourists such as Hard Rock Café and Hooters (both located near Cozumel's Margaritaville) are symbolic of cruise ship culture on the island of Cozumel. Like the cruise ship lines, most of these restaurants are not locally owned but, rather, represent the flavors, sights, sounds, and interiors that are familiar to tourists who hail mainly from the United States. For example, Margaritaville in Cozumel serves "traditional wings," nachos, burgers and fries, chicken tenders, barbeque ribs, and Caesar salads. The interior of the restaurant looks like any TGI Fridays or Applebees in the United States with colorful booths, laminated menus with plenty of pictures, and the sound of popular American music in the background. One TripAdvisor reviewer writes of Margaritaville's food and drink:

> First, we are Buffett fans through and through. Our goal has been to collect a drink glass from every island-based *Margaritaville* restaurant. So we were thrilled to make it to the Cozumel, Mexico location for the first time. The overall menu looks much like the others in the chain, with a few variation [*sic*]. But since we'd had almost a full week of cruise ship food, we were ready for the famous "Cheeseburger in Paradise." While we don't pay lots of attention to the purple USDA stamp at home, apparently the lack of similar oversight can affect hamburger [*sic*.] in our neighbor to the south. We barely recognized the meat. It was 8 oz., but it was tightly machine-pressed, kind of rubbery, and lacking in burger taste. My wife and I both were taken aback and disappointed by it. On the other hand, we were having such a good time with the band, the drinks, and the overall party atmosphere that we ate the fries, a portion of the burgers, and just moved on … We were thoroughly entertained by the two servers walking around on stilts dispensing Blue Curacao shots and a pineapple something or other that burned a little going down but tasted good anyway. (Anonymous 2018)

This review exemplifies tourists' desire to experience what is familiar while traveling internationally. Of course, this desire varies depending on the tourist, including their personality, motivations for travel, income and socio-economic status. In the above example, tourist desires fluctuate between the material, like collecting a drinking glass from multiple destinations, to the more carnal: tasting a piece of beef that reminds them of home. The desire to "let go" is also common among tourists, and as exemplified above, alcohol plays a large role for many vacationers.[4] Although they were not able to taste the memory of home in their cheeseburger, the entertainment the restaurant provided and the taste and consumption of alcohol trumped any residual disappointment this couple may have initially experienced.

Outside the restaurant section of Cozumel's Margaritaville, swings are mounted to an outdoor bar, overlooking an array of inflatable toys used to climb and jump off into the ocean, a large inflatable pyramid and trampoline being the most favored. Employees float around the space, both indoors and outdoors, engaging tourists to "let go," unwind, and party. Young and attractive men and women are the designated "shot givers," there is a man making balloon animals and hats followed by a pair of costumed pirates also administering shots, and an employee who is dressed in a full-body, fluffy shark costume despite the eighty-degree heat. This parade of employees makes its way from table to table, encouraging tourists to drink shots (administered only after a flamboyant and flirtatious display of physical contact between the "shot girl" or "shot boy" with each tourist) while they are taking pictures, making balloon hats, singing, and dancing as well as taking food and drink orders.

The focus of Margaritaville in this section demonstrates how taste influences tourists to seek out experiences, including food and drink, which are familiar to their lives at home. When cruise ships dock, *el malecón* (the esplanade between roadway and sea), which is named Avenida Rafael Melgar (or just "Melgar"), is lined with cruise ship tourists.[5] The busiest restaurants and bars are those that advertise in the cruise line approved maps and shopping guides; they are chains or "entertainment restaurants," and they are close in proximity to the three cruise ship piers. It is rare to see cruise ship tourists even a few blocks inland from Melgar, walking around downtown San Miguel. Any given day a ship is in port each of these international chains are filled with cruise ship tourists, and come nighttime, after the cruise ships leave for their next port of call, each restaurant is deserted or closed completely, preparing for the next day's arrival of cruise ship tourists.

Cozumel's Margaritaville also offers tourists the chance to both experience and viscerally taste a raucous scene of controlled debauchery. An employee of a similarly-themed entertainment restaurant to Margaritaville, located in Puerta Maya cruise ship pier and shopping mall, explains that she earned more tips the wilder the tourists became—that is, stripped of their inhibitions and thus paying more for food, drink, and ultimately a higher tip. It was part of her job description to facilitate a "good time" to cruise ship tourist patrons. Yet, cruise ship tourists in both locations are still in the confines of multinational industry: whether that be inside a US-owned cruise ship terminal and shopping location such as Puerta Maya, or inside the walls of a US-owned company like Margaritaville. Debauchery is controlled in these spaces as taste is familiar and "letting go" only goes as far as ordering another round of drinks, wearing a balloon hat, dancing with a stranger, or stumbling back to the cruise ship. Tourist taste is therefore constructed not by local Cozumeleños but US-owned corporations, which ultimately sanitize the experience and taste of the services and food they provide cruise ship tourists. Tourist taste is also constructed independently of its physical location in the Mexican Caribbean. Taste is constructed to satisfy the tourist's expectation of the familiar but also the economic bottom line of US and other multinational businesses, rather than locally owned options. The ways in which "entertainment restaurants" construct taste is evidence of the industry's purposeful sanitization of experience but also tourists' desire for such a taste, as the two operate in tandem.

Cozumel walking food tour: Challenging the control of tourist taste

The idea of "sanitized taste" can also be contextualized outside of internationally owned chain restaurants and bars. Locally-owned and operated restaurants, bars, cafés, stores, and tours confront "sanitized taste" and, in turn, challenge the ways tourist taste is controlled by the cruise ship industry or multinational businesses that are present in Cozumel. This section describes how a small, locally owned and operated food tour challenges the cruise ship industry, and therefore the construction of sanitized taste, by offering cruise ship and other international tourists the opportunity to experience and taste the island, free from corporate influence.

Julia is an expatriate[6] from the United States who has lived on the island for over five years and works as an entrepreneur and independent tour operator. She owns and operates a multidimensional business catering primarily to international tourists: she specializes in walking food tours around the island, cooks meals for tourists at their resort or hotel as a private chef, and conducts cooking classes for tourists. The relative success of Julia's tour is evident in her TripAdvisor rating, which was designated as "excellent" by 150 out of 153 reviewers; TripAdvisor is one of the most crucial promotion vehicles at Julia's disposal, as well as for other independent operators on the island. Julia describes the difficulties in establishing her business, on TripAdvisor and elsewhere; she has spent countless hours developing and maintaining a website, Facebook account, and TripAdvisor pages specific to each of her businesses. She personally responds to every message posted on Facebook as well as keeping the lines of communication open through email with potential, current, and past tour participants.

There is significant competition among tours and attractions on the island that operate independently from the cruise ship industry. Independent tours like Julia's must reach potential cruise ship and stay-over tourists directly,[7] without the help of all-inclusive resort or cruise ship corporations like Royal Caribbean International or Carnival Corporation & PLC to help market and sell tours. This is made particularly difficult by the fact that the cruise ship industry has its own set of designated tours and attractions that it promotes to passengers. One popular food tour, or "shore excursion," that is operated in Cozumel and endorsed by Carnival Corporation & PLC is "Salsa and Salsa: Mexico's Cooking and Dancing Tour." The price for this shore excursion is $69.99 per adult (13+ years old) and $49.99 per child (4–12 years old). Only cruise ship tourists who are already booked on a Carnival cruise ship may purchase this tour. One of the 387 reviewers of "Salsa and Salsa" describes this excursion on Carnival's website:

> I would highly recommend this fun excursion—learned how to make 6 different salsas. There were enough chips to eat with the salsa, although they could have been refilled more. They were not the best chips either. The cheese quesadillas they served with the salsa were cold, as were the chicken tacquitos [*sic*]. They taught us how to make one margarita on the rocks and then we had unlimited frozen margaritas, which was a nice treat—although they were pretty weak with alcohol. Maybe that wasn't such a bad idea. The two teachers were patient and articulate. They did a great job of trying to teach us to salsa dance following the cooking. The setting is lovely! My biggest complaint was that I thought the excursion was overpriced by about $10. I thought all of the Carnival excursions offered on this trip were WAY TOO EXPENSIVE, based on other excursions I've done. (Carnival Corporation & PLC 2018a)

Expense is certainly a factor for tourists choosing tours at each port of call; however the cruise ship industry assures its tourists that this price is justified based on convenience, quality, confidentiality, and safety. For example, cruise ship tourists often said they were worried about missing the ship if a tour went over its allotted time or there was no prearranged transportation to and from the cruise ship (i.e., "safe" transportation, arranged by the cruise ship, and not via a local taxi or *combi*). Carnival's website alleviates tourist concerns of safety:

> All shore excursions sold through *Carnival* are coordinated with reputable tour operators and include all of the most popular sites of interest. One of the many benefits of booking excursions through *Carnival* is a guarantee that the ship will remain in port until all guests are back onboard. *Carnival* will not be aware of shore excursions that are booked independently. Also, keep in mind that some ports have visa requirements that may prevent you from venturing off on your

> own ... Since all your shore excursions are charged or credited if pre-purchased, to your on board account, you will not have to share your credit card information with anyone you are not familiar with. (Carnival Corporation & PLC 2018b)

Locally owned and operated tours attract a unique type of tourist. First, travel websites such as TripAdvisor attract cruise ship tourists who are curious about the port of call in which they will be disembarking and want information about activities, services, and other leisure opportunities the island provides that their cruise line might not provide. Julia articulates the type of tourist who typically chooses her tours: "They're better tourists. Not the balloon hat, fanny pack, Hawaiian shirt kind ... the tour attracts a different kind of tourist, more adventurous ... who wants to go off the beaten path." Julia consistently compares the "balloon hat tourist," who is often seen inebriated walking around *el malecón*, close to their cruise ship pier, with her sense of a "different kind of tourist." The "balloon hat" tourists of Cozumel disembark on the island for only a few hours. Their main purpose, according to Julia, is to get drunk at either Carlos'n Charlie's [sic], Hooters, Margaritaville, or Señor Frogs—establishments that are not "off the beaten path," located only steps from the piers where cruise ship tourists disembark. Balloon-hat tourists who frequent these establishments wear their balloon creations, often with a drink in hand, as a symbol of their status as cruise ship tourists while they explore the port of call.

In contrast, Julia's conceptualization of a "different kind of tourist" is one that is curious about the island, willing to stray from Melgar and the cruise ship piers, and to seek out new experiences and tastes while traveling. While these kinds of tourists may not be as plentiful or readily accessible, there are benefits to Julia's more natural, unsanitized approach to experiencing the port of call. The type of independent tour offered by Julia allows for flexibility in terms of the services Julia offers and therefore a more personalized experience for adventurous tourists seeking to more meaningfully experience local flavors. Further, Julia has the ability to conduct direct outreach to tourists while remaining independent from the control of the cruise ship companies, which would significantly restrict the type of direct correspondence Julia and her customers engage in. The function of Julia's food tours is to offer a unique experience for tourists who want to learn about contemporary food culture on the island, get "off the beaten path," and show a side of Cozumel the cruise ship industry and other multinational businesses on the island do not provide.

Julia operates a food tour in which she guides tourists around the island to various eateries, street vendors, and food markets that are owned and operated by, and primarily serve, local residents. Tourists on this food tour get a sense of what everyday life is like on the island by interacting directly with locals and buying and eating food that is unique to this part of the world. As food is an integral part of Yucatecan culture (Ayora-Diaz 2012), it is essential for tourists to experience *comida económica*, food that is prepared with fresh ingredients and by hand rather than mass-produced and flash frozen, in order to genuinely experience Yucatán. This is in stark contrast to the spaces manufactured by businesses where cruise ship tourists frequent. Consuming an unsanitized experience, in choosing this type of independent tour, and going "off the beaten path" along with eating food that is not mass-produced, challenges tourists' desire for the familiar and controlled. As noted by a young Canadian couple that opted for Julia's tour over those experiences and spaces highlighted by the cruise line, "Here you see school kids and Mexican families. Just a five-minute car ride and it looks and feels totally different than where the ferry drops you off!"

Models of sanitization and taste: Locally sourced or corporatized?

The cruise ship industry has successfully created an economic arrangement in port of call communities such as Cozumel wherein several aspects of tourists' experience are under its economic control. This outcome, known as vertical integration, is manifested in many all-inclusive and mass tourism contexts, particularly located around the Caribbean Sea (Pattullo 2005). This is accomplished in part by cruise ship companies bundling travel options together under the auspices of consumer ease. Packages often include airline, tour, hotel, and food and drink options that tourists can easily book in advance of their travel. However easily consumers can use and benefit from such a model, vertical integration also limits consumer choice, and therefore taste, by providing consumers with options that are bounded by the limits of a company's own making. For example, on board each cruise ship, while seemingly countless dining options are available and included on most major cruise lines, much of the food that is consumed on board is imported from the United States. This includes those US-based dining franchises now on board select cruise lines, such as Johnny Rockets for burgers and fries, Ben & Jerry's ice cream, and Starbucks coffee. Imported food and US-based franchises allude to consumer choice while on board; however these food options replicate familiar tastes cruise ship tourists experience on land in their everyday. This model further reinforces the distant relationship between multinational and local options for tourists to experience, as profits through packaged tour and accommodations, and therefore vertical integration, are essentially kept away from local hands.

In contrast to sanitized taste present on land in cruise ship port of calls and on board each cruise ship, independently owned businesses catering to international tourists showcase how taste can be locally sourced from the island's tourism industry. As Julia notes, the type of tourists that chooses her tour are different from the cruise ship or other international tourists who frequent the island. We can categorize this tourist as more adventurous, more likely to research local offerings, and more interested in experiencing their travel destination in terms of locally sourced and traditional cuisines. Julia's niche tour and associated tourists confront not only how the island is divided in its economic structure between local and multinational or foreign owned, but also in how they understand what a tourist is or wants. Nash writes that a tourist is "The stranger … a temporary sojourner who does not share the essential qualities of host group life. As a result, interaction between him and the hosts tends to take place on a more general, impersonal level" (1978: 44). Julia's food tour is breaking this pattern of social detachment between host and guest so often seen in the context of mass tourism. As exemplified in Julia's food tour and the type of tourists who choose this experience, impersonality is dislodged in favor of an intimate encounter and taste with the host community, often times in explicit rejection of the sanitized taste most readily available to visiting tourists. This occurs in a compelling and meaningful context, during the sharing of one's table, recipes, and food that is intimately linked to a place's history and culture.

The personal interaction and attention that Julia provides her guests, before, during, and after each tour, sets her businesses apart from competition associated with the cruise ship companies. She can tailor the requests of tourists who choose her tour into their specific idealization of taste (e.g., a tourist seeking vegan options), an approach enabled by correspondence with prospective tour guests before, during, and after their time on the island. The personal connection Julia forges between herself and her guests is distinct from the surrounding mass tourism landscape in Cozumel, which is a clear example of Fordist mass tourism production, a "one-size-fits-all" and easily reproducible model of large-scale tourism (Torres 2002).

Julia and other independent operators in Cozumel combat the corporatization of the island's economic landscape and the corporatocracy of the cruise ship industry by remaining independent from the cruise lines. Smaller tour sizes of more like-minded tourists coupled with personalized communication and attention set independently owned and operated businesses apart from mass tourism's milieu of corporate entities. Taste is not only sanitized by such entities, diluted to fit tourists' desires for the familiar, but is offered and controlled in such a way to limit tourists' access to and desire for less sanitized options in part by never divulging that such options might exist. One reviewer's statement when commenting on Julia's tour, "Don't be afraid," is particularly poignant in this regard. Many cruise ship tourists have revealed they were wary of drinking the water, taking a taxi, or seeing much of the island outside the gates of each cruise ship pier, a dynamic that undoubtedly limits tourists' ability to experience unsanitized taste. As previously described, this is to some degree perpetuated by cruise lines' marketing materials. North American tourists who are fearful of food and drink that is local to Mexico (from ice cubes to *pozole*) are also influenced by anti-Mexican rhetoric that they are exposed to at home. The current anti-Mexican stance that is advanced by the US government builds upon a historical distrust of Mexican things and bodies in general, but Mexican food in particular. The racial slur "beaner" is a prime example of how an important foodstuff has been historically categorized in terms of someone of Mexican descent thought to be other, lesser than, and thus "dirty" when compared to the Anglo-majority population (Arreola 2015). President Trump has built upon these ethnic and racial slurs and stereotypes. He has infamously labeled Mexicans as "rapists" when disparaging Mexican immigrants in a speech announcing his candidacy in June 2015, "They're bringing drugs. They're bringing crime. They're rapists. And some, I assume, are good people." He later tweeted in June 2016, "I love the Mexican people, but Mexico is not our friend. They're killing us at the border and they're killing us on jobs and trade. FIGHT!" Employing such grossly aggressive language against Mexicans with the words "rape," "kill," and "fight," undoubtedly impacts US tourists' associations with Mexico as such rhetoric may breed insecurity for one's safety while abroad. Further, and by extension, this distrust and fear impacts their relationship towards Mexican food, categorizing it as "dirty," "unsafe," and not fit for US consumption.

A vision of sanitized taste as corporately controlled, cold, and aloof, or banal and tacky, is present throughout this chapter. However, to critique *how* one tastes is limiting at best. Sanitized taste offers tourists the familiar but also an opportunity to "let go." The crux of the corporate manipulation, however, is that tourist taste is carefully procured within the limits to one's safety and happiness and the economic bottom-line of each cruise line.

Notes

1 The global cruise ship industry has evolved into only three corporations controlling the global cruise ship market: Carnival Corporation & PLC, Royal Caribbean International, and Star Cruises Group. Star Cruises Group focuses mainly throughout Asia while both Carnival Corporation & PLC and Royal Caribbean International employ itineraries to Canada, Europe, Central and South America, Mexico, and the Caribbean. Carnival Corporation & PLC and Royal Caribbean International are the focus of this research because they are both headquartered in the United States (Miami, Florida) and market directly to North American tourists. Royal Caribbean Cruises Ltd. own and operate the following cruise line companies: Royal Caribbean International, Celebrity Cruises, and Azamara Club Cruises. Carnival Corporation & PLC own and operate the following cruise line companies: Carnival Cruise Lines, Seabourn Cruise Line, Holland America Line, and Princess Cruises. As of June 2014, these two corporations own and operate sixty-three cruise ships in service.

2 Ethnographic data for this chapter was taken from 2011 to 2013 field seasons in Cozumel, Mexico, and on board the Royal Caribbean cruise ship, *Jewel of the Seas*, a five-day cruise which left Tampa, Florida. The itinerary included stops at Cozumel and Costa Maya, which is Royal Caribbean's privately owned port at Mahahual, Mexico, south of Tulum in the Riviera Maya. During these field seasons, I conducted interviews (formal and informal) with fellow cruise ship and international tourists, crew members, tour operators, shop employees (both on and off the ship), expats living on the island, and Cozumeleños.

3 An interviewee stated that the cost of having his locally owned and operated jewelry and gift store listed on the approved marketing materials of the cruise lines initially cost US$10,000 along with a portion of yearly profits which he would be contractually obligated to pay.

4 The role of alcohol in relation to not only taste but as an important experience for many cruise ship tourists is an interesting path of inquiry that should be further developed. In my own research I observed that alcohol contributes to and, it could be argued, is crucial to many tourists' experiences while disembarked and on board cruise ships. Tourists' consumption of alcohol, and the associated attitude of "letting go," are encouraged by servers and other hospitality staff, both on board and off.

5 As many as eight cruise ships can dock on the island at any given time, sometimes all at once, especially during the peak season for cruise ship tourist disembarkations which generally lasts from December to April. This translates to thousands of international tourists disembarking in Cozumel every day.

6 The expatriate community in Cozumel is comprised of permanent residents who are originally from the United States, Canada, and Europe. The twenty expatriates I met are mostly retired and have relocated to Cozumel permanently. They have come to call Cozumel home for various reasons including disdain in raising their children in the United States, more business opportunities on the island, the relaxed "island life," and the warm weather.

7 "The most important criterion to recognize is the differentiation between the 'stay-over visitor' and the cruise ship passenger. The former makes the greatest contribution to the Caribbean's tourism economy, as the stay-over visitor requires accommodation in a hotel, resort, villa, condo, vacation club, timeshare property or—in the case of marinas—a berth for the yacht. The stay-over visitor supports those businesses and investors which have made a significant long term financial commitment in a fixed asset in the Caribbean. In contrast, the cruise ship passenger travels on a ship which is generally only here for the high season, when the local hotels—hopefully—have their profitable months, and nowadays 82 percent of the average cruise ship passenger's discretionary spending is on board" (MacLellan 2013).

References

Anonymous (2018), "Jimmy Buffett's Margaritaville," www.tripadvisor.com/Restaurant_Review-g150809-d1347077-Reviews-Jimmy_Buffett_s_Margaritaville-Cozumel_Yucatan_Peninsula.html (accessed February 13, 2018).

Arreola, C. (2015), "Ethnic Slurs: The History of the Word Beaner," *Latina Magazine*, January 6, 2015, www.latina.com/lifestyle/our-issues/ethnic-slur-beaner-meaning-history (accessed February 13, 2018).

Ayora-Diaz, S. I. (2012), *Foodscapes, Foodfields, and Identities in Yucatán*, Amsterdam: CEDLA, New York: Berghahn.

Carnival Corporation & PLC (2018a), "Salsa & Salsa and Margaritas," www.carnival.com/shore-excursions/cozumel/salsa-and-salsa-and-margaritas-304146 (accessed February 13, 2018).

Carnival Corporation & PLC (2018b), "Shore Excursion FAQs," https://help.carnival.com/app/answers/detail/a_id/2866/kw/shore%20excursion/session/L3RpbWUvMTQ5MjcwNDU4NS9zaWQvZlVLZjFnYzVKOGZRR1paRUxsTjJEdDd3Qk1hSTFTb21qRWtwWnhpZXRJTXZHR0d3cyU3RVRrU2NSZjdodkV3NFhzUXBQSGRZcVAlN0VJVTZDWXBXVjJ4azJkUXpQaDdMWkZTc3JIZWdBYkxXanpoVHJHVV9VbWQ0MTZTdyUyMSUyMQ%3D%3D (accessed February 13, 2018).

Judd, D. and S. Fainstein, eds. (1999), *The Tourist City*, New Haven, CT: Yale University Press.

MacLellan, R. (2013), "Stay-Over Visitors vs. Cruise Ship Passengers: An Important Distinction that the Caribbean Hotel and Tourism Association Forgot," *Barbados Free Press*, June 11, 2013, https://barbadosfreepress.wordpress.com/2013/06/11/stay-over-visitors-vs-cruise-ship-passengers-an-important-distinction-that-the-caribbean-hotel-and-tourism-association-forgot/ (accessed February 13, 2018).

Margaritaville (2018), "About Margaritaville," www.margaritaville.com/about (accessed February 13, 2018).

Nash, D. (1978), "Tourism as a Form of Imperialism," in V. Smith, ed., *Hosts and Guests: The Anthropology of Tourism*, 33–48, Oxford: Blackwell.

Pattullo, P. (2005), *Last Resorts: The Cost of Tourism in the Caribbean*, New York: Monthly Review Press.

Preble, C. (2014), "Imperial Consumption: Cruise Ship Tourism and Cozumel, Mexico," PhD dissertation, State University of New York at Albany, Albany.

Torres, R. (2002), "Cancun's Tourism Development from a Fordist Spectrum of Analysis," *Tourist Studies*, 2 (1): 87–116.

Urry, J. and J. Larsen (2011), *The Tourist Gaze 3.0*, 3rd edn., New York: SAGE Publications.

Postface: Is there Mexican Food? Taste and the Politics of Cultural Identity

Richard Wilk
Indiana University

Authenticity

The chapters in this collection make it clear that there is no such thing as "Mexican food," given a long and complex history; the dramatic environmental, cultural, and culinary diversity of the country; and the vast number of immigrants and sojourners who have imported their own cuisines. But I beg to differ. Within five kilometers of my home in Bloomington, Indiana, I can find plenty of authentic Mexican food, all of it thoroughly familiar to Anglo natives of the region. This includes a place that has been here more than thirty years, selling old-fashioned "plate of tacos, enchiladas, or tamales (or a combination) accompanied by rice and refried beans, all covered with melted cheese." We also have a long established Taco Bell right in the center of town, with an almost infinite number of variations on the theme of tortillas, cheap meat, cheese, and chopped lettuce, and a recent arrival in town full of extremely colorful paintings and carved wood, loud mariachi music, cheap pitchers of margaritas, and buckets of iced Mexican beers, where the fajitas and fish tacos are almost an afterthought. Then we have a new place that makes a determined stab at Mexico City style *tortas* in addition to the standard *chimichangas, quesadillas, enchiladas,* and *burritos.* There is a mobile taco truck run by a family from Puebla where I can get *tacos de cabeza, tripa,* and *lengua* if I get there early enough. And then we have the usual assortment of student burrito places and sit-down chains such as Casa Brava, Chipotle, Qdoba, and Chili's.

Of course, from the point of view of a person born and raised in Mexico, or even a well-traveled tourist, it would be ludicrous to call any of these "Mexican restaurants" except perhaps the taco truck, and even that one has watered down its salsa and otherwise accommodated to local tastes and available ingredients. My point is simply that a national cuisine looks very different from a distance, and like the Belizean food I have written about elsewhere, when a complex cuisine travels it is usually simplified in a process I have likened to "compression." And like Chinese food in the United States, Mexican food has entered at different times and through different routes, with distinct local varieties reflecting both immigrants and the indigenous Hispanic cuisines of the parts of northern Mexico that were appropriated by "manifest destiny." I have also followed Jeffrey Pilcher's "global taco" to places as diverse as Malaysia, Finland, Sweden, Singapore, Belize, and Brazil. And like Pilcher, I can see that the question of authenticity is really beside the point. Taco Bell is so much a part of American everyday dining that it appears on US military bases, and its franchises appear all over east and southeast Asia, northern and southern Europe, Dubai, Australia, and many parts of South America.

Mexican food has therefore become a localized object detached from the nation, in the processes of interaction between migrants, tourists, entrepreneurs, and large multinational corporations. As an unruly beast, it moves throughout the world, popping up in a Singapore mall in the form of "Mexican Turkish Food" (the chain is called Stuff'd, see Stuff'd 2016), in Cincinnati, Ohio, in the form of a thin meat sauce served over spaghetti, or in Finland with a notable absence of corn, chili, or beans. Each place has developed its own variant of Mexican food, which can then become real and authentic to loyal local diners, who might even forget that the food ever had a connection to Mexico. We can liken this process of global diversification to the genetic founder effect in genetics, a form of genetic drift where new populations are founded by only a few migrants that lack the diversity of the founding population. Founder affects, according to evolutionary science, provide both opportunities for rapid evolutionary change and the dangers of a genetic bottleneck, through which bad genes can quickly come to dominate a population.

The implications for cuisine are obvious—some of these new kinds of Mexican food, like Cincinnati chili, rapidly become completely unrecognizable to Mexicans, even though they may become completely authentic in their new habitat. Acquiring a life of their own, they may eventually return to the homeland, as with an American-style Chinese restaurant opening in Beijing (though Taco Bell has still failed in Mexico after two attempts). As Vassallo-Oby points out in

Chapter 12, we can look at this process as one of homogenization and corporate simplification (what George Ritter calls McDonaldization), but it can also be described as a new form of diversification, at a global level where each nation has a number of culinary stereotypes. We could also see it as a kind of two-level tourist marketing, where mass tourism consumers are fed a comfortable and easily swallowed form of culinary diversity, and the high-cultural-capital consumers can seek out more challenging, specialized, localized, and "scholarized" variants. The first needs a tour guide or a hotel concierge, while the second needs a food Sherpa, a specialized food tour, or a visit to someone's home to learn to make an exotic dish. Rather than looking at one as authentic and the other as fake, we are better off treating both of them as motivating different forms of authentification, targeted to different audiences.

Food scholars have moved onward to recognize that in every instance authenticity is constructed in different ways, often by contending interests who take familiar positions. Should we adapt old dishes to new ingredients, change them to match the tastes of younger generations, or should we revitalize old traditions by returning to rural roots and forgotten varietals? Is fusion food okay if we maintain the authentic "spirit" of the indigenous kitchen? How can we reconcile very different kinds of food that are being presented as local? What do we do about food traditions rooted in a form of rural poverty that people are desperate to escape? When do we cross the line between respectful borrowing and the violence of cultural appropriation? The authors of this book avoid these traps and adopt instead a critical awareness of the many interests involved in defining cuisines.

Nationalism

The shadow of UNESCO is cast over every one of these chapters. In our time commodification has steadily advanced, encompassing more and more of religion, morality, health, art, and human creativity. National character and citizenship have conventionally been seen as inherent cultural and biological qualities that rest in nature (climate, soil, water, genes) and the transcendent power of the state rather than on the marketplace. A nation chooses its own official flag, flower, bird, anthem, and coat of arms; and we can now add a cuisine, ingredients and/or dish as a materialized form of "intangible cultural heritage." Previous constructs like "French cuisine" could be freely adopted, adapted, and marketed, with only a general understanding that they in some way represented the accumulated history and knowledge of the place. With UNESCO certification, however, they become a form of property that can be possessed and defended, thus acquiring legal status as objects rather than subjective experiences. A national cuisine is now more like a brand, which has a measurable monetary value that economists call brand equity. This equity in turn can become a major asset for a corporation, bought and sold; for companies such as Coca-Cola, brand equity is a larger asset than office buildings.

But is a cuisine really like other forms of property? Unlike a fragment of poetry, a recipe cannot be copyrighted or given trademark protection. No two moles are ever exactly alike, and even something as prosaic and common as a *tortilla* has a lot of variation in texture, flavor, etc. The closer we get to specifying the exact nature of a dish, a meal, or a style of cooking, the more problematic it becomes, and this is why so many scholars have found fault with the UNESCO certification process. But the marketplace does not care about our misgivings and quibbles, and no matter how much we argue about the constructed nature of authenticity, it still has tremendous market value. We could even argue that in an age of mass production and global trade, authenticity has become the master commodity (e.g., Heller 2003) that we ignore at our peril.

One of our practical problems in addressing cuisine as a commodity is the uncertain level of specification. For example, the *milpa* and several specific crops are included in the definition of UNESCO's traditional Mexican cuisine. Eating together as a family group is part of the Mediterranean diet, specific events are marked by a French gastronomic meal, while Croatia focuses on gingerbread as a defining dish, and Turkey identifies with a specific method of making coffee. Of course, all of these are elements of cuisine, but where are the boundaries if everything is connected to food? When we compare cuisines we are too often comparing not just apples and oranges, but elements as unlike as wine and tablecloths, birthdays and types of wheat.

The diversity of what we mean by "food" is reflected in this book. Some of the authors address particular dishes like *salbutes* and tamales, while others are concerned with ingredients like corn and chili peppers. But we can also address culinary diversity by looking at timescales, for example, the things that are appropriate to eat for breakfast and other times of day, or at the time of life, as in baby food or teenage food. Other chapters address the timescale at the longer duration of annual cycles of festivals and their appropriate dishes (e.g., Vargas-Cetina's Chapter 7), historical time scales (e.g., Fernández-Zarza & López-Moreno, Chapter 4), and even evolutionary changes in the nature of plant and animal foodstuffs (e.g., Fernández Souza, Chapter 1). At a smaller scale, we can think about the sequencing and co-occurrence of foodstuffs in snacks and meals, like Cohen and Kershenovich Schuster (Chapter 8) asking how *chapulines* should be eaten, and Ayora-Diaz (Chapter 6) asserting that Yucatecans avoid the combination of black beans with *cochinita pibil*, while learning to accept *mucbil pollos* stuffed with cheese and ham.

Territory

Beyond time, the distribution of food in space is the other major theme in this volume (as in contemporary food studies in general). How does food define places and locations, create and transcend boundaries, and establish different kinds of geographical relationships and connections? Unfortunately, most food scholars have inadvertently adopted an archaic "nested box" model of location, built on the work of German anthropogeographers in the nineteenth century. Like modern bureaucratic states, this model requires each person to live in a household, each household to belong to a community, and each community to belong to a physical territory, defined by ethnicity, landform, and/or language.

This way of counting, classifying, and controlling citizens and their movements has been useful to demographers, census-takers, and state bureaucracies, but as the authors in this collection show, it is not an accurate way to map and understand cuisine and culinary culture (and it is becoming more problematic for nations, a problem that cannot be solved by building a wall). Instead we are better off thinking of cuisines or foodways as "fuzzy sets" held together by their relationships to a prototype, rather than being contained by boundaries or borders (Lakoff 1987; also Wilk 2010). So, while Yucatec or Oaxacan foods may be quite distinct, and we may identify them by their putative location in a particular Mexican region, these cuisines are actually in perpetual motion, with new dishes, spices, and combinations being absorbed and other things being exported abroad. As Brulotte (Chapter 5) shows, the very identity of Oaxacan mezcal is motivated by connoisseurship in the USA and elsewhere, while Pérez (Chapter 9) demonstrates that Oaxacan food is developing in part as a performance of locality for a tourist audience.

A nested box approach to food territory is also inherently hierarchical, taking the form of a stepped pyramid, with the vast diversity of individual and family recipes and practices at the bottom, and national cuisine at the top. Again, this is very useful for writing tourist brochures and marketing, since it is a form of simplification with which everyone is completely familiar. The pyramid is a good way to rank things, and it corresponds easily with old anthropological models of segmentary opposition. But for a practicing food scholar, the pyramidal hierarchy is inherently problematic because there are so many disconnects between levels (is Yucatec food part of Mexican food, or does it have a separate identity?). Furthermore, most individuals partake of more than one cuisine, some of which are more important to their personal identity than others. We are all multilingual when it comes to taste, and we all learn what linguists call code switching as we adapt our taste and expectation to the context of each conversation.

Sorting foods into cultural boxes is still a common mode among popular food writers and individuals in search of authenticity, but the authors of this book are scrupulous in avoiding the fallacy that some forms of food are authentically indigenous and others are alien or fake. A good example is Pilcher's discussion (Chapter 10) of the competition between beer and *pulque* among Mexican drinkers, and while pulque is indeed pre-Hispanic and indigenous, they both have an established history in Mexico. There are indeed "authentic" Mexican beers, just as there have always been many authentic and local recipes and varieties of pulque. This awareness has also led these authors to avoid falling into the lazy practice of depicting change as a simple story of the "replacement of the local by a foreign invader." This is another favorite theme of popular food writing, going back to the time of the Romans (and probably long before), part of the eternally codependent relationship between the local/indigenous and the foreign/global (Wilk 2006).

Concluding remarks

Although this book begins with the question "is there Mexican food?" each chapter finds a different way to approach the issue of how social groups define and use their foods, an inherently better question than asking if food defines a social group or nation. Once again, I think an analogy with population genetics might be helpful. Think of national cuisine as the equivalent of "race." The untrained citizen looks at the diversity of human features and tries to sort them into discrete groups, and scientific racists then look for biological or genetic boundary markers to make those groups seem natural. In reality, most forms of human genetic variation, like skin color, vary in frequency along clines. And when we measure the frequency of many different genes, we find that they do not fall into discrete clumps or populations that are genetically isolated. Despite what the popular gene self-testing kits say in their advertising, there are no pure ancestral populations that can be unmixed to tell you who you really are. This is because there is always more genetic variation within a population than we can find between populations. Yes, there are traits and genetic variations that can be traced back through time to particular sources, but there are no races in the sense of pure genetically isolated human groups. There are certainly no groups that can be genetically defined on the basis of superficial characteristics like eye or skin color.

Human beings as a species mate beyond social boundaries, just as they eat beyond social boundaries. To extend the analogy, we should not expect to find pure unmixed ancestral cuisines—everything is a mixture, a creole, a blend. Social and legal boundaries around nations or ethnic groups surely have an effect, but we should never expect to find entirely different cuisines separated by natural or political dividing lines. To extend the analogy to race, the concept of a national or local cuisine is perfectly real in the sense that many people believe in it, but good scholarship requires us to recognize the difference between actual practice and what people believe (etic and emic to use the language of anthropology). The difference between the two is often the most fruitful place to begin research. So, while I began this essay by asserting that Mexican food does not exist, I am very happy to end it by saying exactly the opposite. Mexican food is alive and well as long as millions of people believe in it, as long as restaurants and corporations sell it, and as long as it remains a topic for our endless curiosity as scholars.

References

Heller, M. (2003), "Globalization, the New Economy, and the Commodification of Language and Identity," *Journal of Sociolinguistics*, 7 (4): 473–492.

Lakoff, G. (1987), *Women, Fire, and Dangerous Things: What Categories Reveal about the Mind*, Chicago: University of Chicago Press.

Stuff'd (2016), "Concept," http://stuffed.com.sg/ (accessed December 10, 2017).

Wilk, R. (2006), *Home Cooking in the Global Village: Caribbean Food from Buccaneers to Ecotourists*, Oxford: Berg.

Wilk, R. (2010), "Consumption Embedded in Culture and Language: Implications for Finding Sustainability," *Sustainability: Science, Practice and Policy*, 6 (2): 1–11.

Index